Servo Motor
and Motion Control
Using Digital
Signal Processors

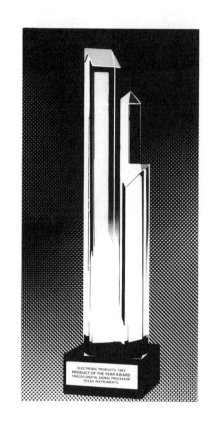

**Prentice Hall and Texas Instruments
Digital Signal Processing Series**

Servo Motor and Motion Control Using Digital Signal Processors

Yasuhiko Dote, Ph.D.

Professor, Electronics Engineering Department,
Muroran Institute of Technology, Japan

Prentice Hall   Englewood Cliffs, New Jersey 07632

Library of Congress Cataloging-in-Publication Data

Dote, Yasuhiko
   Servo motor and motion control using digital signal processors /
by Yasuhiko Dote.
      p.   cm.
   Includes bibliographical references.
   ISBN 0-13-807025-3
   1. Incremental motion control. 2. Digital control systems.
I. Title.
TJ216.D68 1990
629.8'3--dc20   90-31293
                                         CIP

Editorial/production supervision
  and interior design: *Jacqueline A. Jeglinski*
Cover design: *Lundgren Graphics, Ltd.*
Manufacturing buyer: *Margaret Rizzi*

Published by Prentice-Hall, Inc.
A division of Simon & Schuster
Englewood Cliffs, New Jersey 07632

The publisher offers discounts on this book when ordered
in bulk quantities. For more information, write:
        Special Sales/College Marketing
        Prentice-Hall, Inc.
        College Technical and Reference Division
        Englewood Cliffs, NJ 07632

If your diskette is defective or damaged in transit, return it directly to Prentice Hall at the address below
for a no-charge replacement within 90 days of the date of purchase. Mail the defective diskette together
with your name and address.
        Prentice Hall
        Attention: Ryan Colby
        College Operations
        Englewood Cliffs, NJ 07632

The author and publisher of this book have used their best efforts in preparing this book and software.
These efforts include the development, research, and testing of the theories and programs to determine
their effectiveness. The author and publisher make no warranty of any kind, expressed or implied, with
regard to these programs or the documentation contained in this book. The author, publisher, and Texas
Instruments Incorporated shall not be liable in any event for incidental or consequential damages in
connection with, or arising out of, the furnishing, performance, or use of these programs.

Printed in the United States of America
10 9 8 7 6 5 4 3 2 1

ISBN 0-13-807025-3

Prentice-Hall International (UK) Limited, *London*
Prentice-Hall of Australia Pty. Limited, *Sydney*
Prentice-Hall Canada Inc., *Toronto*
Prentice-Hall Hispanoamericana, S.A., *Mexico*
Prentice-Hall of India Private Limited, *New Delhi*
Prentice-Hall of Japan, Inc., *Tokyo*
Simon & Schuster Asia Pte. Ltd., *Singapore*
Editora Prentice-Hall do Brasil, Ltda., *Rio de Janeiro*

Contents

Section III System Implementation 158

Preface

The rapid and revolutionary progress in power electronics and microelectronics in recent years has made it possible to apply modern control technology—which has gone through evolution during the last three decades—to the area of servo motor and motion control. Currently, microprocessor-based controllers are invariably used in most applications of drive electronics. Although digital signal processors (DSPs) have been developed primarily for application in the field of communication, recently they have been used in digital servo motor and motion control because of their suitable architecture and fast computation capability. The uses of DSPs has permitted the increasingly stringent performance requirements and fast, efficient, and accurate control of servo motor and motion control systems. This technology can provide high productivity and extensive product quality in the production line and is the basis for modern industrial automation. The purpose of this book is to review this interdisciplinary technology.

Section I in this book presents Background, intended for engineers with a Bachelor's degree in Electrical/Mechanical Engineering who are not familiar with servo motor and motion control techniques. This section is also ideal for use as classroom text.

Section II describes Digital Control and Signal Processing, and is for practicing engineers who are seeking advancement in knowledge of motion control.

Section III gives System Implementation, which was written for practicing engineers who are interested in the usage of DSPs. This section, if desired, can be used as a text for actual DSP experiment. A general overview of DSP servo motor, motion control, and guide to use this book is introduced in Chapter 1.

In this text, the mathematical approach is severely curtailed in favor of description with numerous charts, graphs, and figures that provide actual illustrations of using DSPs in servo motor and motion control. Reprints from current manufacturers' literature are used extensively for this purpose. References are given in each chapter, especially chapters 5 and 6, for more detailed design information on each algorithm. Three engineering books and college textbooks often emphasize theoretical aspects and lag behind the practices. The gap in digital motion control is especially wide because of rapid advances in technology. This book particularly intends to remove this gap.

It is impossible to prepare this type of manuscript without cooperation from many people. I am grateful to Professor C. Slivinsky of the University of Missouri—Columbia, U.S.A., and Professor K. Ohnishi of the Keio University, Japan, for providing the pertinent material for chapters 4 and 7. Thanks are due to Mr. J. Usami, Dr. K. S. Lin, and other staff members of Texas Instruments for encouragement in writing this book and contributions in same. Appreciation is also extended to many graduate students at the Muroran Institute of Technology, Japan, especially to Mr. J. R. Timm and Mr. A. S. M. Gerard for experimentation, typing the manuscript, and drawing the figures and diagrams. Finally, I am thankful to Dr. B. K. Bose at the Power Electronics Applications Center, University of Tennessee, U.S.A., for going through the manuscript carefully, and to Ms. C. A. Williams for typing the final manuscript.

Yasuhiko Dote

Servo Motor
and Motion Control
Using Digital
Signal Processors

Chapter 1

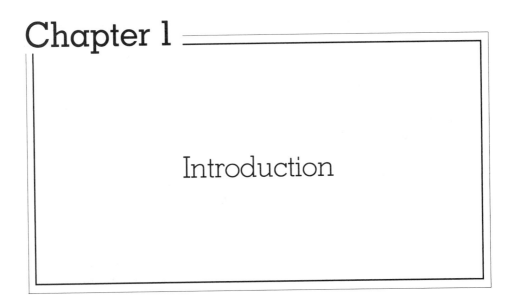

Introduction

1.0 INTRODUCTION

In the field of factory automation (FA), FA equipment such as Numerical Control (NC) machine tools, industrial robots, and like machinery are used to save energy and promote automation, which allows for high productivity and also produces quality products. The aforementioned FA equipment is becoming more technical with the application of servo motor and motion control technologies.

Servo motor and motion control technologies are based on mechatronics engineering (the replacement, combination, and fusion of electronics and mechanical engineering, which is also interdisciplinary). Due to recent, remarkable progress in power electronics and microelectronics, more advanced servo motor and motion control is now available. Although it had been developed primarily for application in the field of communication, the digital signal processor (DSP) is being used as a controller and sensor signal processor because of fast computational capability and suitable architecture.

This book mainly describes the practical realization of servo motor and motion control using the TMS 320C25 for practicing engineers who wish to obtain high productivity on the production line and to produce quality products as well. The TMS 320 family has recently established itself as the industry's standard DSP.

1.1 DIGITAL SERVO MOTOR CONTROL (DSMC) SYSTEMS*

The servo motor is an important component of an actuator. Servos are now light-weight, compact, easily integrated, efficient, controllable, and nearly mainte-nance-free. Easy servo motor maintenance is preferable, especially in locations such as unmanned factories where a great number of servo motors are employed. Brushless servo motors are a big advantage in such applications.

A current passing through the armature coil of a motor is one of the alter-nating sine, square, trapezoidal, or other waveforms. Mechanical commutator switches (brushes) are usually installed on motors to generate alternating currents. Instead of this, alternating current can be generated by external semiconductor switching circuits; hence, the brushless servo motor. Brushless servo motors have the following main advantages:

- higher maximum speed and more capacity
- work in less favorable surroundings under adverse conditions
- lower maintenance costs and less noise

There is also greater freedom in planning the shape with brushless motors. Motors can be made more compact and there is easier integration with main equipment.

Brushless servo motors are classified as stepping motors [Induction Syn-chronous Motor (ISM), Induction Reluctance Motor (IRM)], dc brushless motors or permanent magnets [Synchronous Motor (SM)], and vector-controlled motors [Induction Motor (IM)]. Their schematics are shown in Fig. 1.1.

Figure 1.2 shows the optimal capacity ranges for each motor type. These figures show many factors such as mechanism of torque generation, operational efficiency, cooling, control devices, control techniques, ease of manufacture, and profitability.

The components necessary for the control of a servo motor are: the main motor, angular, angular velocity, current, voltage, magnetic flux, temperature sensors, and a semiconductor power converter (power amplifier) including various analog and digital ICs for triggering control. In addition to these, a small-sized motor driving gear with a position and speed sensor is mounted on the motor shaft, and a digital controller (DSP) is also included. The entire system is shown in Fig. 1.3. In the signal generation part, motor voltage, current, flux, and fre-quency are controlled in order to obtain accurate and immediate torque response. Motion controls (position, speed, and force control) are obtained in the control part.

As described, the control system has to be considered as a single unit where the objective is the control of a high-performance servo motor. Compromising the performance of some components can be allowed if the performance of the whole servo motor driving system can be improved. The driving system needs to

* See reference [1.1].

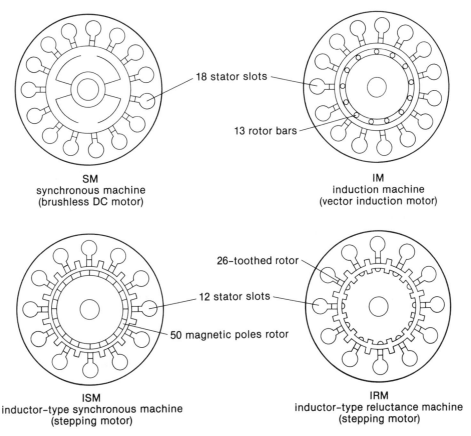

Figure 1.1 Various structures of brushless servo motors

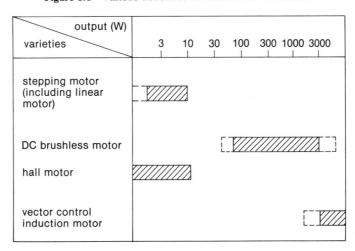

Figure 1.2 Optimal capacity ranges of brushless servo motors

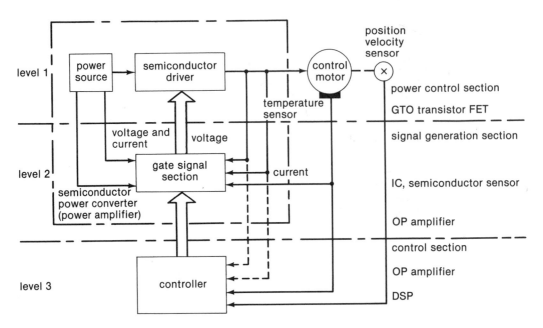

Figure 1.3 Driving gear of brushless servo motors

be designed and applied as one unit body to make the best use of the features of the semiconductor power converter, the detectors, and the main frame of the motor. Each component should be designed to support the other's performance.

In evaluating a servo motor driving system, the control performance, speed ratio, speed of the motor, applicable torque range, and acceleration performance are stressed. The deciding factors of control performance and the speed ratio are the power rate and the accuracy of detectors in both the motor's main frame and the power converter. Detectors must be able to find rotational positions for position control and to detect speed for response rate and/or speed control.

The speed signal is produced by processing the position signal in most digital control cases; consequently, a resolver or an encoder is used for that purpose. Therefore, the encoder and the small-sized motor are sometimes combined into a single unit (the encoder motor) in order to make the system smaller. This method gives multiple functions to one motor. However, if the necessary resolution cannot be obtained by a position sensor, then a brushless (brush) tachogenerator must be used.

A resolver trades off high sampling rate for the high resolution that is lacking in the encoder. The encoder's pickup is stationary, but the pickup of a resolver rotates with excitation frequency which can be controlled freely. These give the resolver much more flexibility as a position sensor. Detectors can have resolutions as low in gauge as 1.2 millionths of one revolution, making very smooth rotation control possible.

Semconductor power converters have been developed as the main part of power electronics techniques. Bipolar transistors and MOSFETS (Metal Oxide Semiconductor Field Effect Transistors) having high power handling capability along with high-speed switching characteristics are produced commercially. Their production technique is being improved all over the world. Variable frequency and PWM (Pulse Width Modulation) inverters and converters have been produced using these high-performance components. Some of these components have working frequencies of 10–30 kHz.

Highly advanced IC production techniques make possible analog and digital ICs, custom LSIs, and one-chip microcomputers for use as gate signal generators. These components allow for good control performance and reliability. IC techniques are also applied in the design of high speed/power semiconductor devices.

Software has been developed that makes it possible to control the current, voltage, and frequency of a servo motor with a considerable degree of freedom (highly efficient control techniques are used). This control is performed by feedback—or feedforward control of current, voltage, position, and speed.

Brushless motors have, due to orthogonally controllable magnetic flux and current, the control performance at least equal to that of dc motors. In addition, tasks impossible with brush motors have been made possible by the brushless servo motor. Namely, they are sine wave output current control to reduce torque ripple, high-speed operation to make a motor smaller and lighter, improvement of the efficiency of converters to save energy, and equivalent field-weakening control. Furthermore, semiconductor converters have sequence control abilities such as starting and stopping, malfunction detection, self-diagnosis, and self-protection.

Direct motor power transmission to a targeted object (without gear trains) is preferable. This is performed simply by use of direct drive from a linear motor. The control principles of that method are similar to other principles. The controller shown at the bottom of Fig. 1.3 is usually used for performing regulation and tracking with respect to position and speed. A DSP controller is being introduced into the design of controllers.

Since modern control theory includes classical control theory, it has become possible to design a complete compensator having larger control capability than the traditionally used (i.e., for higher disturbance suppression) PID controller (partial compensator).

Furthermore, accurate and reliable control, insensitive to disturbances and all types of parameter variations, as well as fast control, can be designed by applying modern control theory and be put into practical use. Varieties of LSI, VLSI microprocessors and DSPs play important roles in realizing this goal.

There is a demand in mechatronics to shorten the time involved from design development to production as much as possible, because product life cycle is decreasing. One way to meet this demand is to use CAD (Computer Aided Design) in engineering semiconductor power converters and controllers. It has been suggested to use a method applying a two-terminal circuit and a computer to exactly simulate a semiconductor power converter. In one case, a PID controller was designed by a high-speed personal computer using practical convolution model method, based entirely on specifications (nonphysical model).

All of these methods require great computational effort, so they are performed by high-speed, digital signal processors (DSP) combined with personal computers. DSP will be used more, not only as digital controllers and signal processors, but also as analyzers and design equipment.

As described above, electronic techniques are necessary for hardware that is required for the control of servo motors. Extended modern control theory's applicability to software is coming into existence because of the improvement of hardware.

The operational principles of brushless servo motors will be explained by means of a mechanical system utilizing the concept of analogy between electric systems and mechanical systems.

The operation (control) principles of a dc brushless or brush motor and a vector control induction motor will be explained with reference to Fig. 1.4 (see reference [1.2]). Figure 1.4(a) shows the brush motor's operational principles, and Fig. 1.4(b) shows those of the brushless motor. Figure 1.4(a) shows that the disc is driven by a tangential force f_T working on pin p. Pin p is moved radially by an internal force f_R of the slot. The radial movement of the pin is slow because there is a considerable amount of friction. Friction is dependent on the speed (a permanent magnet field has constant R). The following analogy is formed between this type of dc motor (an electromagnetic system) and a mechanical system.

$$f_T = \text{armature current}$$

$$R = \text{main field flux}$$

$$f_R = \text{field current}$$

$$Rf_T = \text{electric torque}$$

$$Rw = \text{electromotive force (EMF)}$$

$$Rwf_T = \text{electric power}$$

It is easy to control the position of disc (p), speed (w), and torque.

The reason for that is that only force f_T needs to be controlled, since R in Fig. 1.4(a) is constant (a permanent magnet being used to make the field of dc brushless servo motors). The length of a slot represents the saturation of field and the maximum EMF.

In a dc brush motor, the commutator selects a slot to apply f_T, depending on the angle of rotation of R (the field's position). In a dc brushless servo motor, the function of the commutator of a dc brush motor is performed by a pole sensor and semiconductor power converter.

In a vector control induction motor, pin p in Fig. 1.4(b) is driven by the

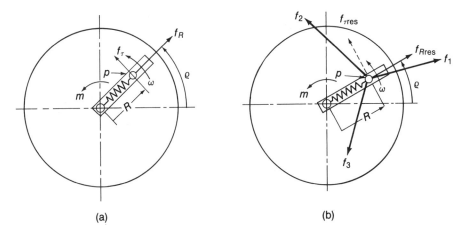

Figure 1.4 Operational principles of each variety of brushless servo motor by means of analogy to a mechanical system. (a) Brushless servo motor, stepping motor; (b) vector control induction motor.

forces f_1, f_2, and f_3, each having its definite direction. Forces f_1, f_2, and f_3 have to be simultaneously adjusted in order to keep the radius R constant.

As just described, it is difficult to control a moving coil motor. The use of one power converter, however, permits simultaneous control of f_{Tres} and f_{Res}. The principle of vector control is independent control of f_{Tres} and f_{Res}. For that reason, f_1, f_2, and f_3 are calculated from the desired f_{Tres} and f_{Res} by means of coordinate transformation. The instantaneous position of pin p, and field flux R, have to be detected at the time of coordinate transformation. R changes in the case of an ac machine and its angle of field flux rotation is different than that of the rotor's.

Therefore, the problem of vector control is to ascertain how to determine R and the instantaneous position of p in the slot. They are estimated using the same model as this one. However, problems arise due to errors in the estimated values. These errors are caused by changes in the coefficient of friction in the slot due to temperature changes. It has been reported that the problem has been solved by calculating the coefficient of friction by online computation using a microprocessor which corrects the values.

The above discussion has been written about brushless servo motors; this book focuses primarily on dc brushless servo motors.

Brushless dc Servo Motor Drive System

Brushless servo motors have been developed in the industrial fields requiring large capacities along with the development of applied technology of thyristors. Since Siemens Co. of West Germany announced the production of Hall motors of frac-

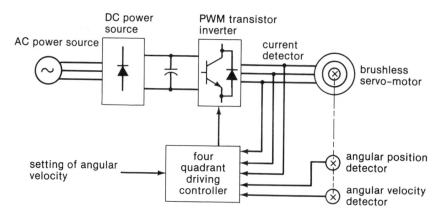

Figure 1.5 Driving system of a brushless servo motor

tional horsepower, the development of brushless servo motors has been promoted all over the world.

The application of brushless servo motors became attractive due to several factors: reduction of the price of power transistors, establishment of the technique of current control of PWM inverters, development of permanent magnetic materials, development of varieties of highly accurate detectors, and manufacture of these components in a compact form. In this way, brushless servo motors were equipped with the delicacy of dc motors and the strong structure of ac motors. In addition, they were completely freed from the output power reduction due to the commutator and from maintenance complexity. It is no exaggeration to say that brushless servo motors created a new phase in the development of control motors.

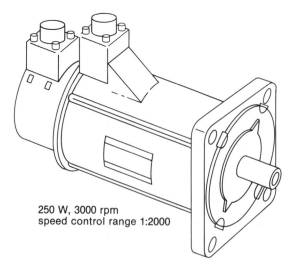

250 W, 3000 rpm
speed control range 1:2000

Figure 1.6 An example of a brushless servo motor

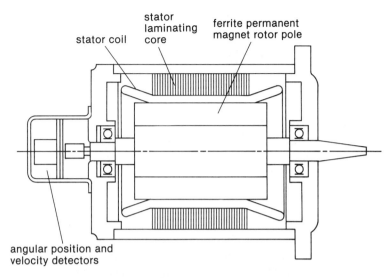

Figure 1.7 Internal structure of brushless servo motor

As explained in the previous section, the princple of brushless servo motors is that the function of the commutator of the dc brush motor is substituted by a pole sensor and a semiconductor power converter. The generated torque is proportional to the product of current and field flux which is orthogonal (as with ac motors).

Figure 1.5 shows the general configuration of a brushless servo motor. Figure 1.6 shows the appearance of a brushless servo motor of 250 W, 100 V, four pole, 3000 rpm, three-phase stator coil, and ferrite permanent magnet rotor flux field.

A rotor position detector and a speed detector are mounted on the motor shaft. A thermal detector is installed on the stator to detect overload operation.

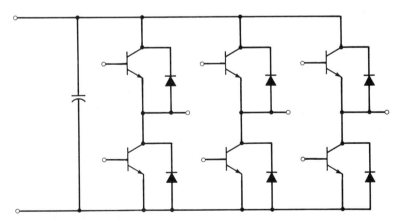

Figure 1.8 Main circuit of current control transistor PMW inverter

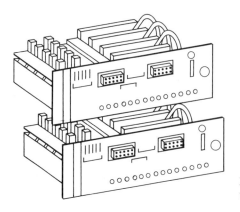

Figure 1.9 An example of a current control transistor PWM inverter

Most of the loss in the motor takes place in the stator overload, where it is easily detected; furthermore, the stator housing can be designed for easy cooling.

Ferrite is chosen because of its low price, relatively high performance, and particularly low sensitivity to demagnetization, which make it a good choice of magnetic material for the rotor pole. The armature current does not influence the distribution of magnetic flux when the motor is driven under normal load conditions because the linear torque-speed curve can be measured. In addition, the material should be examined to determine whether it can withstand field demagnetization.

The magnetic circuit has large reluctance, partly because a permanent mag-

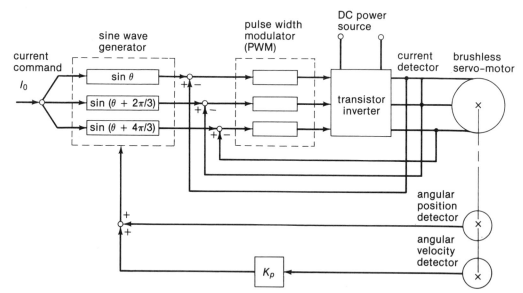

Figure 1.10 Control block diagram of the driving system of a brushless servo motor (excluding speed control loop)

net is used and there is a comparatively large magnetic gap. Therefore, the primary circuit has small inductance and it is possible to pass a large amount of current in a short time frame; this contributes to a fast response. The internal structure of the motor is shown in Fig. 1.7. The main circuit and the appearance of a current control transistor PWM inverter are shown in Figs. 1.8 and 1.9, respectively.

The current waveform can be made close to a sine wave by means of PWM (Pulse Width Modulation) because this inverter uses phase current feedback. With that, energy wasted in the motor, torque ripple, and transistor peak current can all be minimized.

Figure 1.10 shows the control block diagram of the driving system of a brushless servo motor. Three-phase sine waves are generated by the signal produced at the angular position detector attached to the rotor. They are multiplied by the current command value and compared with each of the feedback three-phase currents. If the motor current is larger than the command value, the inverter switches itself in the direction that limits the current. If the motor current is smaller than the command value, the inverter switches itself in the direction that increases

TABLE 1.1   AN EXAMPLE OF THE CHARACTERISTICS OF THREE BRUSHLESS SERVO MOTORS

Item	Model	ASM-061M	ASM-121M	ASM-252M
rated output	W	60	120	250
rated torque	kg·cm	1.95	3.90	8.12
rated speed	rpm	3000	3000	3000
rated voltage	V(AC)	37	41	103
rated current	A(AC)	1.6	2.7	2.2
maximum torque	kg·cm	9.75	19.5	40.6
maximum speed	rpm	4000	4000	4000
power rate	kW/s	0.85	2.16	3.72
torque constant	kg·cm/A	0.95	1.13	2.88
mechanical time constant	msec	14.3	8.94	7.24
electric time constant	msec	1.85	2.33	3.72
EMF constant	mV/rpm	11.2	13.2	33.3
rotor inertia (J)	kg·cm·s^2	0.44×10^{-3}	0.69×10^{-3}	1.74×10^{-3}
armature resistance	Ω	2.21	1.23	2.54
protection structure			totally enclosed	
insulation type			type B	
environ- tempera- mental ture	°C		0 ~ 40	
condition humidity	%		90 RH or less (no dew condensation)	
weight	kg	2.4	3.0	5.2

Source: Shibaura Engineering Works Co. Ltd.

the current. The speed signal is fed back to estimate the future angular signal, which reduces the time delay of the system.

The structure and architecture of motors with a capacity of 0.1 kW–4.8 kW have been described in this section. In addition, skew must be introduced to eliminate cogging. Then, motors and detectors need to be improved to reduce torque ripple. The dead zone of semiconductor power converters should be eliminated, and the current offset in current controllers has to be cleared.

These types of motors are presently being manufactured, in power outputs ranging from several watts to those of several kilowatts. Examples of performance are shown in Table 1.1 and Table 1.2.

These types of motors permit simple maintenance and have long life and high reliability, because they have no brushes and no commutators. Moreover, they can be used even in environments with dust, explosive, or corrosive materials and still maintain efficiency. In the case of power failure, they can apply dynamic braking. Since they are easy to balance, they have a low vibration level. They have a linear torque-current characteristic and, being highly accurate, they have superior current waveform, high-speed response, and a wide speed control range. Therefore, they are optimum for the control of position, speed, and force.

An example of motor current waveforms during a step response of speed is shown in Fig. 1.11(a). Figure 1.11(b) shows a Bode diagram of the entire system.

TABLE 1.2 AN EXAMPLE OF CHARACTERISTICS OF BRUSHLESS SERVO DRIVERS

Item	Model	ADM-061A	ADM-121A	ADM-252A
applied motor	W	60	120	250
voltage of main circuit	V(DC)	140	140	280
voltage of control circuit	V(AC)	100/200/220	100/200/220	100/200/220
maximum output current	A	6	8	8
maximum input current	A	3.7	6.3	4.5
speed control range			1000:1	
speed regulation load charge			(0–100%) 0.1% or below	
voltage fluctuation			(±10%) 0.1% or below	
temperature change			(±25°C) 0.5% or below	
control method			transistor PWM produces a sinusoidal wave	
speed feedback			semiabsolute encoder	
speed command voltage	V(DC)		10/rated speed	
command input impedance	kΩ		10	
environmental temp*	°C		0–50	
condition hmd**	%		90RH or less (no dew condition)	
weight	kg		3.5	

Source: Shibaura Engineering Works Co. Ltd.

temp*: temperature

hmd**: humidity

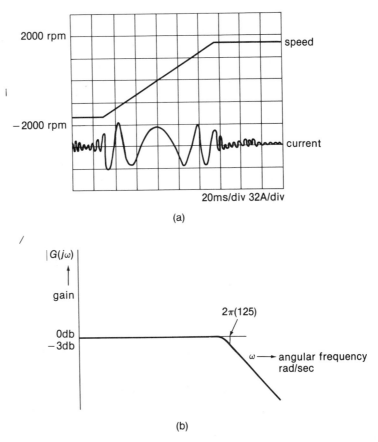

Figure 1.11 An example of motor current waveforms and Bode diagram of the whole system. (a) Speed response to step command and primary phase current waveform. (b) Bode diagram of the speed control system.

High accuracy and speed response can be expected up to the corner frequency of 124 Hz. The time constant of the current loop is about 2 ms.

As shown in Table 1.1 and Table 1.2, brushless servo motors are a little inferior to dc brush motors in terms of efficiency and size when their capacities are very small. However, they are control motors of high performance because they have less vibration and greater longevity compared with dc brush motors.

1.2 DSP CONTROLLERS

The recent rapid and revolutionary progress in microelectronics has made it possible to allow a control design apply modern control theory (a well-developed discipline during the last few decades) to the implementation of digital servo motor

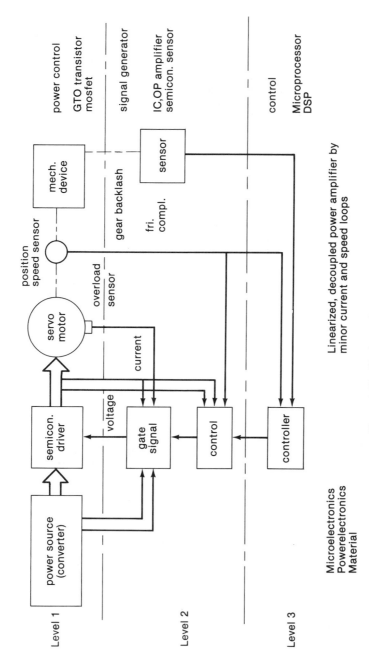

Figure 1.12 Hardware for control system

control (DSMC). Microprocessors, such as the TMS 320C25 digital signal processors, are increasingly being used to implement algorithms for the control of robust and optimal fast dynamic systems. The major factors contributing to this trend are the availability and lower costs of digital signal processors.

Future typical applications of DSP are in the control of fast, complex mechanical devices in which servo motors are used as actuators. In these applications, the required control bandwidth varies from 100 Hz up to several kHz, and their like sampling frequencies must be considerably higher than these.

Software, hardware, and algorithms for designing and implementing a software servo motor control system by using the fast microprocessor, the Texas Instrument TMS 320C25 digital signal processor (DSP), are described in this section. This section also highlights the practical realization of a PI brushless servo motor speed control algorithm using the TMS 320C25.

By using a brushless servo drive system, motion control (position, velocity, and force control) can be achieved. The hardware for the control system is shown in Fig. 1.12, with DSPs used as controllers.

Figure 1.13 shows the block diagram of a digital control system. The digital controller consists of a computer, or a microprocessor/microcontroller, which implements a control algorithm in real time. The D/A converter transforms the digital output of the microcomputer $u(n)$ into a suitable analog signal $u(t)$ (i.e., voltage, current) for the motor. The output $u(t)$ of the D/A, which may need power amplification, will drive the motor to the desired or reference position/velocity, $r(n)$. The output of the plant or the motor $y(t)$ which may be position or velocity measured by a sensor like a tachometer, potentiometer, or a shaft encoder is converted into a digital signal $y(n)$ (if necessary) by the A/D. The feedback signal is subtracted from the reference signal $r(n)$, to create an error signal $e(n)$. The error signal $e(n)$ is used by the controller to issue the corresponding control action $u(n)$.

If the controller is implemented by executing computer programs (a set of difference equations written in assembly language), this can be called software servo motor control (SSMC).

The performance of most mechanical systems is not very consistent and varies depending on external environmental conditions. During operation, both

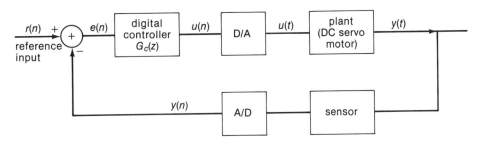

Figure 1.13 Digital control system

inertia and required torque vary. High-performance operation cannot be expected due to the disturbances caused by these variations.

In the SSMC, how the system operates is determined by the computer software. In other words, various kinds of control strategies can be served by one control system, and a variable-structured controller can be easily designed. The strategies of the controller can be changed or updated during operation in order to obtain higher control performance. This is performed by writing a program, storing it in memory, getting the sensed variables, and then calculating the optimal control.

For example, a program for a PI controller is as follows:

$$U(i) = Kp(Ri - Xi) + \sum_{j=0}^{i} Ki(Rj - Xj) \qquad (1.1)$$

where $U(i) \triangleq$ input to D/A converter (control)
$Kp, Ki \triangleq$ constants
$R \triangleq$ command
$X \triangleq$ controlled variable
$j \triangleq$ sampling instant
if Kp = not 0, Ki = not 0, then the controller is a PI controller.
if Kp = 0, Ki = not 0, then the controller gives integral control action.

Furthermore, Kp and Ki can be adjusted (adapted) according to the mechanical behavior without changing hardware. In SSMC, digital signal (from sensors) processing can be performed by writing programs. Therefore, the advantages of adapting SSM are as follows:

1. Without changing hardware, a control strategy can be easily changed.
2. By executing software, the undesired mechanical behaviors can be canceled or compensated.
3. The undesirable performance of a sensor and a controller can be cancelled or compensated. Thus, total servo performance can be improved.
4. It is easy to construct a large-scale control system (a distributed control system) since one servo system owns its computer and it communicates with a host computer or another computer.
5. The constructed system has a reduced size, weight, and power with low cost.
6. The system gains greater reliability, maintainability, and testability.
7. The system gains increased noise immunity.
8. Sophisticated control algorithms for higher control performance can be realized.

Thus, one has a robust controller with which accurate and fast control may be planned, and easily designed and operated by software. The purpose of de-

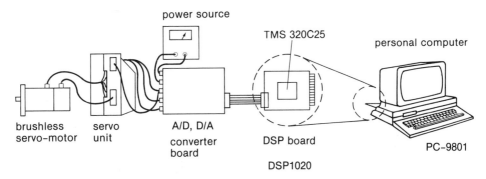

power source

TMS 320C25

personal computer

brushless
servo-motor

servo
unit

A/D, D/A
converter
board

DSP board

DSP1020

PC-9801

Figure 1.14   Schematic diagram of overall system

signing a robust controller is to obtain high-performance (accurate and fast) control action.

Examples for DSP Controller

The following is an example of brushless servo motor speed control and implementation with a digital signal processor (TMS 320C25). Figure 1.14 shows a schematic diagram of the overall control system. A proportional and integral control algorithm given in Eq. (1.1) is implemented with a TMS 320C25 (DSP).

The PI controller is applied to speed control for a brushless dc servo motor.

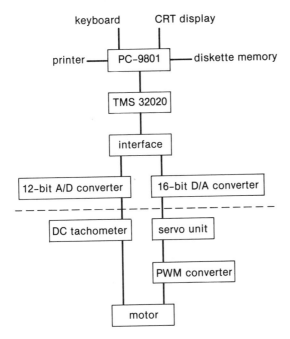

keyboard   CRT display

printer — PC-9801 — diskette memory

TMS 32020

interface

12-bit A/D converter   16-bit D/A converter

DC tachometer   servo unit

PWM converter

motor

Figure 1.15   Hardware details using DSP (TMS 320C25)

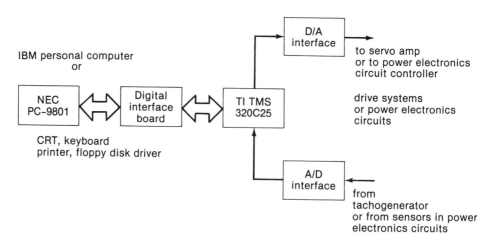

Figure 1.16 Configuration of DSP controller

The heart of the system is a DSP (Texas Instruments TMS 320C25) with a PC-9801, which is based on an INTEL 8086 microprocessor. The system includes a keyboard input device, a CRT display, a diskette memory, its driver, and a printer (which are used for developing the program and for storing and displaying the measured values). A 12-bit D/A converter provides sufficient accuracy for this control purpose. A tachometer is used as the sensor. A PWM transistor converter-fed brushless dc servo motor is used. Hardware is shown in Fig. 1.15. A 100 W

```
*
*        cntrol data make
*
CONST   LAC    MODEL     *err=model-omega
        SUB    OMEGA
        SACL   ERR
*
        LT     ERR
        MPY    IGAIN     *(16.16)=err(16.0)*igain(0.16)
        PAC              *i control(integ)=err*igain
        ADDH   INTEG     *             +integ
        ADDS   DINTEG    *             +dinteg
*
        SACH   INTEG
        SACL   DINTEG
*
        MPY    PGAIN     *p control(32.0)=err(16.0)*pgain(16.0)
        PAC
        ADD    INTEG     *control=i control+p control
        SACL   CNTRL
```

Figure 1.17 Control program written by assembly language

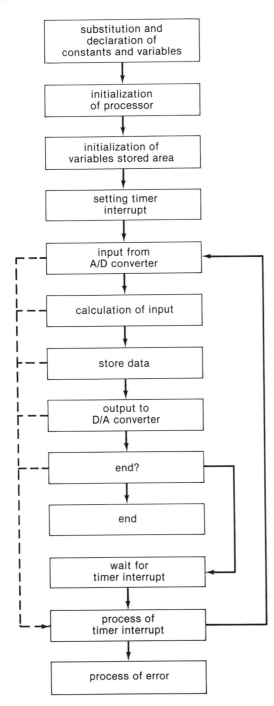

Figure 1.18 Flowchart for whole program

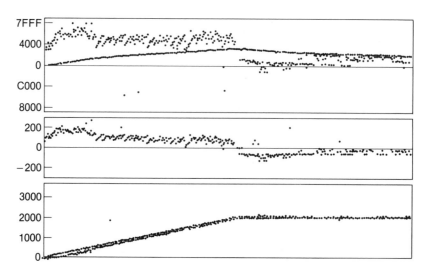

Figure 1.19   Experimental results

Shibaura servo pack supplying a dc motor with a fixed field is used here. A small brushless dc tachometer provides an analog speed feedback signal.

This digital control development system, shown in Fig. 1.16, consists of a NEC 9801 microprocessor (INTEL 8086) and a digital interface board, a DSP and A/D, D/A interface boards. In operation, the DSP program is downloaded to the TMS 320C25 from the PC-9801 for execution. The TMS 320C25 has stored initialization program and data memory areas. Each block of data is transferred from the host computer to the DSP. Experimental data are brought from the DSP to the PC-9801, then displayed on the CRT display or on the printer.

Figure 1.17 shows an assembly language program. Figure 1.18 shows a general flow chart. The control results are shown in Fig. 1.19.

Currently, the DSP is being applied to motion control all over the world: as a notch filter for suppressing mechanical oscillation, a nonlinear compensator in a hydraulic system, and active vehicle suspension controller, a controller for a robot, a parameter identifier, a state estimator, a digital sensor signal processor, a spectrum analyzer for CAD tool, inverter control, and other applications as well.

1.3 BOOK ORGANIZATION AND HOW TO USE THIS BOOK

This book covers a fairly wide range, since the described technology is based on the combination, fusion, and replacement of mechanical and electronic engineerings; it is an interdisciplinary area (called mechatronics).

However, the author would like to highlight novel control and signal processing algorithms that make it possible to eliminate a gap between theory and practice, since they are robust and practical and they make it easy to determine controller (or digital filter) parameters. Section II (chapters 5 and 6) is intended for practical engineers seeking information on advanced motion control.

A brushless servo motor drive has been developed by using mechatronics engineering, and its use is preferred due to its maintenance-free and high control performance nature. The author believes that engineers (readers) studying this servo drive can understand any other type of servo drives. Section I (chapters 2 through 4) covers a fundamental brushless servo control system. This section is provided for engineers with a B.S. degree in electrical or mechanical engineering who are not familiar with servo motor and motion control; for those engineers, this section can be used as a textbook.

Section III (chapters 7 through 9) is provided for engineers who are interested in the usage of a DSP. System implementation issues of control and signal algorithms with a DSP (TMS 320) are described. This section may be used as a textbook for experiments and appendices.

This book is organized as follows:

- Chapter 1 introduces the outline of a DSP servo control system.
- Chapter 2 covers digital servo motor control components in detail.
- Chapter 3 illustrates the control theory used in this book.
- Chapter 4 describes motion control where mechanical impedance control concept and mechanical oscillation suppression are noted.
- Chapter 5 covers digital robust control algorithms, with which accurate and fast servo motor and motion control can be expected. Controllers and sensors are mutually complementary when they are used as compensators.
- Chapter 6 deals with robust digital signal processing (including observers) with a DSP to obtain high-performance but low-cost sensors.
- Chapter 7 presents DSP compensator implementation issues.
- Chapter 8 describes how to develop and write servo motor and motion control programs using an assembly language.
- Chapter 9 covers the practical implementation results performed in the author's laboratory of position tracking, speed regulation, and grasping force controls. This chapter is believed to be useful for practicing engineers who plan to use DSPs to control equipment in their plants and to manufacture products. Appendices are provided for experiments.

BIBLIOGRAPHY

[1.1] R. Gabriel and W. Leonhard, "Microprocessor Control of Induction Motors," Proceedings of the International Semiconductors Converter Conference, Orlando, U.S.A., May 1982.

[1.2] Y. Dote and S. Kinoshita, "Brushless Servo Motor: Its Fundamentals and Applications," Denshi Sogo Shuppa, Tokyo, Japan, June 1985.

Chapter 2

Digital
Servo Motor Control
Components

2.0 INTRODUCTION

A brushless servo motor has been developed by using interdisciplinary mechatronics engineering. A mechanical switching device (commutator) is replaced by
electronic circuits and controllers resulting in a maintenance-free and high-performance servo drive.

By studying this chapter, another type of servo motor drive can be understood, since this chapter contains various kinds of technologies used in the servo
motor drive development.

2.1 ACTUATORS: SERVO MOTORS AND SERVO DRIVERS
WITH CONTROL CIRCUITS*

The brushless servo motor is an ac motor in all respects and it is in fact called
the "ac servo motor." Combined with a dedicated control device, however, the
performance of the brushless servo motor is found to be equal or superior to the
efficiency of the high-performance dc servo motor.

The rotational speed of the dc servo motor is varied generally by changing

* See references [2.1] and [2.2].

the voltage applied to the armature. As the armature voltage is nearly proportional to the rotational speed, the latter can be slowed down as much as desired by lowering the former.

On the other hand, the rotational speed of the ac motor is varied generally by changing the frequency; however, the frequency has its limits of variation. A wide range of speed variation, which is a feature of servo motors, cannot be obtained only by using a simple inverter.

In this section, the techniques of giving superior controllability to ac motors will be studied with reference to the principles of dc motors.

Principles of dc Motors

The principles of the dc motor are shown in Fig. 2.1 in a simple manner. When the current flows in the electric conductor after traveling through the brushes and the commutator in the magnetic field generated by permanent magnets N and S, torque is generated in the direction of the arrow shown in Fig. 2.1 in accordance with the Fleming left-hand rule. When the rotor is turned by about 90 degrees, the direction of the current is reversed by the action of the commutator. Thus, the rotor continues to rotate.

Let us examine this series of operations in detail. As the rotor is energized and rotated from the position illustrated in Fig. 2.1, the torque gradually decreases and becomes zero when the rotor is rotated by 90 degrees. The rotor does not stop rotating at that position but continues to turn because of the rotor intertia. When it rotates by a little more than 90 degrees, commutation takes place and the torque begins to increase gradually.

The torque generated by the motor shown in Fig. 2.1 varies widely. For practical use, however, each of the motors has dozens of commutator segments, so even slight degrees of rotation cause commutation. These motors are designed so that they may be used constantly at their maximum torque.

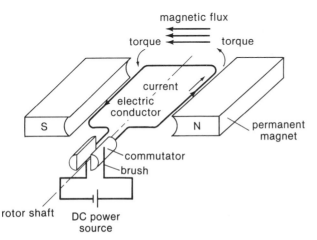

Figure 2.1 Basic principle of DC motors

The line of magnetic force intersects the direction of current at right angles in a dc motor. This enables the motor to always gain a stable torque proportional to the current. For that purpose, rectifiers (brushes and a commutator) are indispensable.

Principles of ac Motors

In the motor shown in Fig. 2.2, a slip ring is used instead of a commutator. When the motor is energized to make brush A positive and brush B negative, torque is generated to turn the rotor as in the case of dc motors. However, the rotor ceases to turn later because of the lack of a commutator in the motor. Consequently, the direction of the current should be inverted at the power source at the right time.

Conversely speaking, a power source of alternating current makes the rotor turn continuously at the rotational speed corresponding to the applied frequency. From this, it follows that the ac motor is rotated by synchronizing the rotation to the frequency of the power source, and the brushless motor is rotated by reversing the polarity of the power source according to the changing rotor position. Both types of motors have the same structure.

The motor shown in Fig. 2.2 is equipped with brushes and a slip ring, but the brushes can be eliminated in the structure of the motor, as shown in Fig. 2.3. Both types of motors, however, rotate on the basis of the same principle. The former is called the revolving-armature type and the latter is called the revolving-field type. Most of the brushless servo motors are of the revolving-field type.

Principles of Brushless Servo Motors

The brushless servo motor lacks the rectifiers which the dc motor has, and has a device for making the current flow to fit the rotor position by controlling the

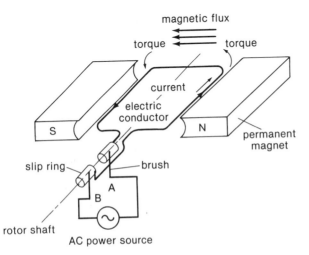

Figure 2.2 Basic principle of AC motors (1)

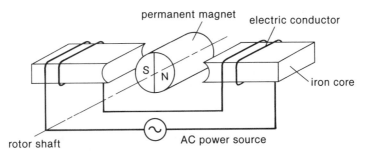

Figure 2.3 Basic principle of AC motors (2)

power source. In the dc motor, torque variation is reduced by increasing the number of commutator segments. On the other hand, in the brushless motor, torque variation is reduced by making the coil three-phase and by transforming the current of each phase into sine wave. Figure 2.4(a) and Fig. 2.4(b) are cross-sectional views of a three-phase synchronous motor, with U^+, U^-, V^+, V^-, W^+, and W^- indicating the beginning or the end of the coil of each phase.

When a motor is energized by three-phase alternating currents, as shown in Fig. 2.4(c), only phase U is positive at point A, while phases V and W are both negative. Therefore, the direction of the current of each coil is as shown in Fig. 2.4(a) and the composite vector of the magnetic flux induced by the current is generated in the direction from N toward S. If there is a rotor field intersecting the magnetic flux at right angles at that time, torque is generated to revolve the rotor clockwise due to the repulsive and attractive forces between the magnets.

At point B, magnetic flux is generated more clockwise by 60 degrees. It then follows, from the above discussions, that a continuous rotating field can be obtained by making three-phase currents to flow in the stator coil. If the sine wave phase and the rotational position can be made to meet constantly at right angles, it becomes possible to make a highly efficient motor of smooth torque without using brushes.

The sections above have also covered what kind of current is appropriate for a brushless motor from the viewpoints of structures and principles of brushless motors. The following sections describe the control circuit necessary for outputting the current waveform of brushless motors.

Control Circuit of Brushless Servo Motors

In order to control a brushless servo motor, the control device has to undertake the task of making the magnetic flux perpendicular to the current. This should essentially be performed by the motor, in addition to the task of controlling the voltage applied to the armature by the dc servo motor. The components necessary for the circuit of a brushless servo motor will be examined by comparing them with those of a dc servo motor.

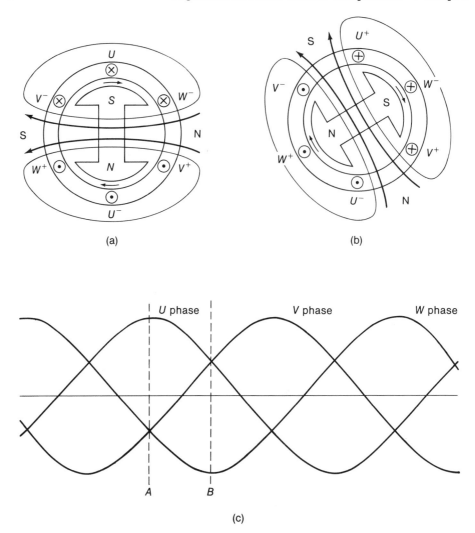

(a)

(b)

(c)

Figure 2.4 Principle of a rotating field

The detailed explanation about the sensor attached to brushless servo motors will be presented in section 2.2. An absolute encoder is used for the control circuit presented in this chapter because its operating principles are easy to understand.

The basic system diagram of the dc servo motor is shown in Fig. 2.5, and that of the brushless servo motor appears in Fig. 2.6. The phase to be controlled is single-phase in the dc motor and three-phase in the brushless motor. Furthermore, the brushless motor has a rotor position detector, a sine wave generation circuit, dc–SIN conversion circuit, a speed detector, and so on. However, the dc motor has none of these.

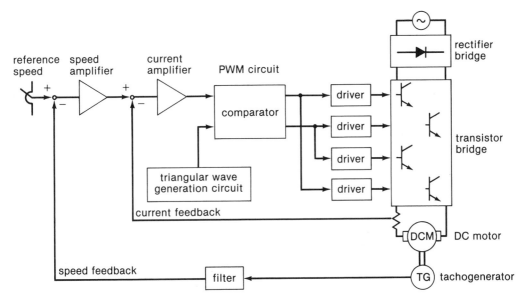

Figure 2.5 Block diagram of the system of DC motors

In the following sections, the functions of those circuits peculiar to the brushless servo motor are examined.

Rotor Position Detector

As previously described, the position of magnets should be known accurately in order to assure that the direction of current and the magnetic flux meet at right angles.

The rotor position detector circuit is used for receiving a rotor position signal from the encoder and converting it to a form that can be read by the sine wave generation circuit which follows the rotor position detector. If the encoder is 8-bit absolute, a code signal generated by dividing one rotation by 256 is sent to the rotor position detector circuit from the encoder. The code is converted into pure binary number as shown in Fig. 2.7. After that, when each binary digit is completely assigned, the generated signal is extracted.

Sine Wave Generation Circuit

This circuit is for generating sine waves according to the rotor position. It is basically composed of ROM (Read Only Memory). ROM is widely used as a memory cell of computers. Its structure is discussed below.

As shown in Figs. 2.7 to 2.9, the necessary data, each of which corresponds to its address, are written on a ROM. Next, when the binary numbers designating the addresses are fed to the address bus connected to the ROM, the data cor-

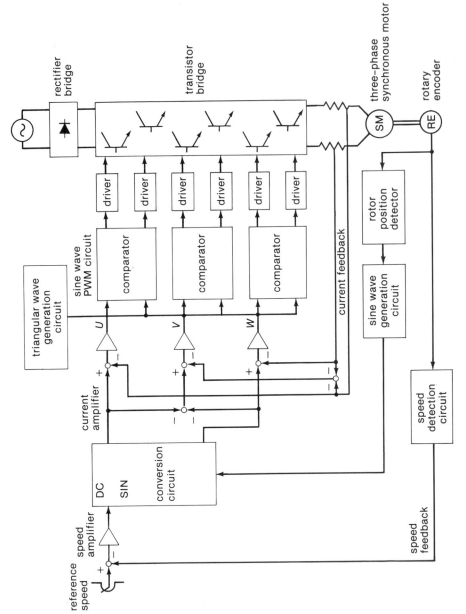

Figure 2.6 Block diagram of the system of brushless servo motors

28

binary digit signal \ rotation of rotor	1/256	2/256	3/256		254/256	255/256	1
2^7	0	0	0		1	1	0
2^6	0	0	0		1	1	0
2^5	0	0	0		1	1	0
2^4	0	0	0		1	1	0
2^3	0	0	0		1	1	0
2^2	0	0	0		1	1	0
2^1	0	1	1		1	1	0
2^0	1	0	1		0	1	0

Figure 2.7   Condition of each binary digit to the angle of rotation of the rotor

responding to the addresses are conveyed close to the data bus. Then the data is sent to the data bus when read signals are fed to the ROM.

By making use of this characteristic, the pattern of the sine wave is memorized by one cycle per one rotation in a bipolar motor, and by one cycle one half rotation in a tetrapolar motor. As the brushless servo motor in this case is a three-phase motor, it should have three phases with phase differences of 120 degrees from one another. In practice, phase V can be estimated by a simple analog operation through the equation $V = -(U + V)$. Therefore, only phases U and W have to be memorized by ROM.

Figure 2.10 and Fig. 2.11 show the values of the data calculated by a computer and written on the ROM, where the addresses of one cycle are in locations 00H-FFH (hexadecimal number), the maximum value of amplitude is in location

address	contents
0	data 0
1	data 1
2	data 2
3	data 3
⋮	⋮
⋮	⋮
254	data 254
255	data 255

Figure 2.8   Contents of ROM

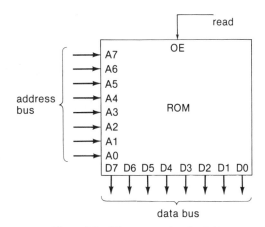

Figure 2.9   Wire connection for ROM

	0	1	2	3	4	5	6	7	8	9	A	B	C	D	E	F
0000	80	83	86	89	8C	8F	92	95	98	9B	9E	A1	A4	A7	AA	AD
0010	B0	B3	B6	B9	BB	BE	C1	C3	C6	C9	CB	CE	D0	D2	D5	D7
0020	D9	DB	DE	E0	E2	E4	E6	E7	E9	EB	EC	EE	F0	F1	F2	F4
0030	F5	F6	F7	F8	F9	FA	FB	FB	FC	FD	FD	FE	FE	FE	FE	FE
0040	FF	FE	FE	FE	FE	FE	FD	FD	FC	FB	FB	FA	F9	F8	F7	F6
0050	F5	F4	F2	F1	F0	EE	EC	EB	E9	E7	E6	E4	E2	E0	DE	DB
0060	D9	D7	D5	D2	D0	CE	CB	C9	C6	C3	C1	BE	BB	B9	B6	B3
0070	B0	AD	AA	A7	A4	A1	9E	9B	98	95	92	8F	8C	89	86	83
0080	7F	7C	79	76	73	70	6D	6A	67	64	61	5E	5B	58	55	52
0090	4F	4C	49	46	44	41	3E	3C	39	36	34	31	2F	2D	2A	28
00A0	26	24	21	1F	1D	1B	19	18	16	14	13	11	0F	0E	0D	0B
00B0	0A	09	08	07	06	05	04	04	03	02	02	01	01	01	01	01
00C0	01	01	01	01	01	01	02	02	03	04	04	05	06	07	08	09
00D0	0A	0B	0D	0E	0F	11	13	14	16	18	19	1B	1D	1F	21	24
00E0	26	28	2A	2D	2F	31	34	36	39	3C	3E	41	44	46	49	4C
00F0	4F	52	55	58	58	5E	61	64	67	6A	6D	70	73	76	79	7C

U

Figure 2.10 Data damp list of sine wave of phase U

	0	1	2	3	4	5	6	7	8	9	A	B	C	D	E	F
0100	EE	EC	EB	E9	E7	E6	E4	E2	E0	DE	DB	D9	D7	D5	D2	D0
0110	CE	CB	C9	C6	C3	C1	BE	BB	B9	B6	B3	B0	AD	AA	A7	A4
0120	A1	9E	9B	98	95	92	8F	8C	89	86	83	7F	7C	79	76	73
0130	70	6D	6A	67	64	61	5E	5B	58	55	52	4F	4C	49	46	44
0140	41	3E	3C	39	36	34	31	2F	2D	2A	28	26	24	21	1F	1D
0150	1B	19	18	16	14	13	11	0F	0E	0D	0B	0A	09	08	07	06
0160	05	04	04	03	02	02	01	01	01	01	01	01	01	01	01	01
0170	01	02	02	03	04	04	05	06	07	08	09	0A	0B	0D	0E	0F
0180	11	13	14	16	18	19	1B	1D	1F	21	24	26	28	2A	2D	2F
0190	31	34	36	39	3C	3E	41	44	46	49	4C	4F	52	55	58	5B
01A0	5E	61	64	67	6A	6D	70	73	76	79	7C	80	83	86	89	8C
01B0	8F	92	95	98	9B	9E	A1	A4	A7	AA	AD	B0	B3	B6	B9	BB
01C0	BE	C1	C3	C6	C9	CB	CE	D0	D2	D5	D7	D9	DB	DE	E0	E2
01D0	E4	E6	E7	E9	EB	EC	EE	F0	F1	F2	F4	F5	F6	F7	F8	F9
01E0	FA	FB	FB	FC	FD	FD	FE	FE	FE	FE	FE	FF	FE	FE	FE	FE
01F0	FE	FD	FD	FC	FB	FB	FA	F9	F8	F7	F6	F5	F4	F2	F1	F0

Figure 2.11 Data damp list of sine wave of phase *W*

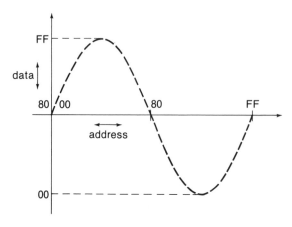

Figure 2.12 Waveform of phase U after converted into analog

FFH, and the maximum value of amplitude is in 00H. Figure 2.10 is for phase U and Fig. 2.11 is for phase W.

Figure 2.12 and Fig. 2.13 show the data indicated by means of graphics display that are equivalent to the analog output of the circuit.

DC–SIN Conversion Circuit

By means of the sine wave generation circuit, two-phase sine waves synchronized with the rotor positon are generated. The sine waves, however, are expressed by varying the amplitude from −1 to +1 through 0, then storing them in 00H-FFH. Hence, these factors have to be converted into practically needed current values.

Therefore, in this circuit, the sine wave reference current is estimated by multiplying the reference current, which is the output of the speed amplifier, by the amplitude factor of the sine wave. As the speed reference signals are fed as direct currents, both in the dc servo motor and the brushless servo motor, the

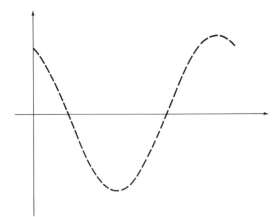

Figure 2.13 Waveform of phase W after converted into analog

speed feedback signal to be compared with them should also be a direct current. Accordingly, the speed amplifier output, which is the result of the comparison, is also a direct current.

In the case of the dc servo motor, the speed amplifier output can be used as the reference current without any modification, because the current of the motor to be controlled is a direct current. In the case of the brushless servo motor, the speed amplifier output has to be converted and used as the reference current, as shown in Fig. 2.14(a). This circuit consists of the combination of a D/A converter, which changes digital signals at the output of the sine wave generation circuit into analog signals, and a multiplier as shown in Fig. 2.14(b).

Sine Wave PWM Circuit

The aim of the brushless servo motor is to make sine wave currents flow in the motor, so it is ideal for a brushless servo motor to have the output of the current amplifier of the sine waves applied directly to the motor after amplifying the power. However, amplification of the sine waves is not practical because that

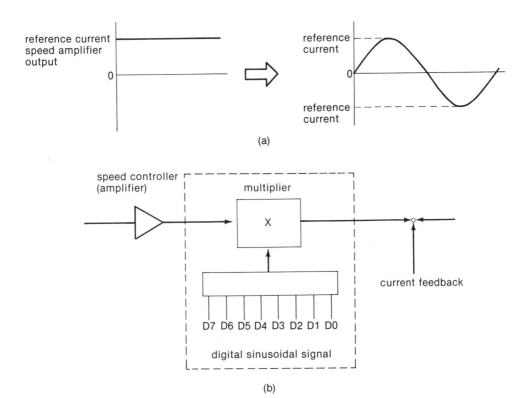

Figure 2.14   (a) Objective of DC–SIN conversion. (b) DC to sinusoidal conversion circuit.

means using a power transistor in the proportional region. This makes it difficult to solve the problem of high temperature due to power loss. Consequently, the power loss is reduced by switching the power transistor. This method is called PWM (Pulse Width Modulation).

In this method, the current of a motor is converted into a controlled pulse width that is proportional to the amplitude of the sine wave, so that it may become a sine wave on the average. Figure 2.15 shows the principles of the method. A triangular carrier wave oscillating with constant frequency and amplitude, and the sine wave output from the current amplifier, are compared by a comparator.

As shown in the figure, pulses of unequal widths are output by extracting the portions where the values of the sine wave exceed those of the carrier wave. The duty ratio of the pulse width is increased or decreased, centering around 50 percent by the sine wave and modulated to make a sine wave on the average, because the inverter output is 0 V when the duty ratio is 50 percent. It is important to decide the method of selecting the oscillating frequency of the chopping wave.

As the carrier frequency equals the switching frequency of the power transistor, it increases the switching loss proportionally as it is made higher, and it reduces the speed of response of the servo motor as it is made lower. Furthermore, ripples appear more frequently, and the torque change and the core losses are increased.

Generally, a carrier frequency of 1–3 kHz is selected when the inverter consists of bipolar transistors, and that of 5–20 kHz when it consists of FETs. The current ripples developed at these levels of frequency make the iron core of

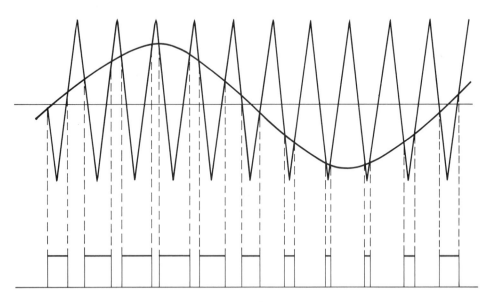

Figure 2.15 Principle of sine wave PWM

the motor vibrate. This generates unpleasant noises when the frequency is within audio range.

In order to solve this problem, the carrier frequency is made 16 kHz or more by using FETs. Another means to solve the problem is to prevent the generation of noises by molding the iron core and the coil of a motor into one body.

Speed Detector

The dc servo motor often utilizes a dc tachogenerator (TG) as a speed detector. However, as the TG has a brush, brushing and maintenance of the motor should be performed simultaneously when the TG is used in a dc servo motor. If we want to make use of the merits of a servo motor, then TG cannot be used. Therefore, some kind of sensor is used to detect the rotor position as well as the speed of the brushless servo motor.

In this chapter, it has been assumed that a brushless servo motor has a rotary encoder as a sensor, so speed detection by using an encoder will be examined. Encoders are classifed as absolute or incremental. The absolute type encoder is obviously necessary, when it is used after the power source is activated.

It is also possible to use the pulses developed from the track of the absolute encoder because the necessary condition for speed detection is to get a pulse train synchronized with the rotation of the motor, as it is also necessary to determine the direction of rotation at the same time. An encoder outputting a two-phase pulse train, with each phase having the same period, and a phase difference of 90 degrees between the phases is required.

The F/V converter is generally used to obtain direct current speed signals from the two-phase pulse train. The following functions are required of a F/V converter when it is used in a servo motor:

1. In order to have a fast response, the time constant of the smoothing circuit has to be reduced.
2. The judgment of normal rotation and reverse rotation has to be performed accurately for each pulse.

Figure 2.16 shows an example of a speed detector satisfying the above conditions.

Both the leading edge and the trailing edge of the waves of the pulse trains of phase A and phase B, developed from the encoder, are adjusted by the synchronous circuit, such that the waveforms are synchronized with the clock pulses. The waveforms are delayed by one clock cycle by the delay circuit.

Synchronized two-phase pulses and the two-phase pulses delayed by one clock cycle are fed to the decoder circuit. The frequency of the clock pulses is usually set somewhat higher than the maximum frequency of the encoder pulses. In the decoder circuit, pulses of quadruple frequency having a pulse width of one clock period are input separately according to the logical condition of each pulse.

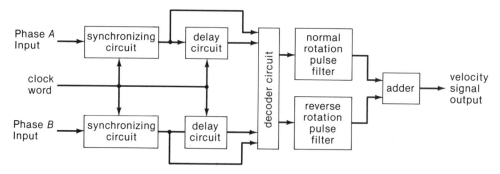

Figure 2.16 Output waveform from encoder

DC signals proportional to the speed are obtained as the frequency of each pulse (keeping its given area) changes with the change in the speed of rotation of the motor. This is shown in Fig. 2.17.

It is impossible, in theory, to remove ripples completely by a smoothing circuit because a smoothing circuit originally smooths square waves. Hence, the time constant should be set to about $\frac{1}{10}$ of the rise time (step response) of the motor, and the pulse number of the encoder should be determined so that ripples are within the allowable value.

In the F/V method, the pulse number of an encoder tends to increase when the motor rotates at extremely low speeds. Accordingly, the output of an encoder is made an approximate sine wave and speed signals are extracted from its differential output. This is a method designed to make a motor rotate at extremely low speed with a small pulse number.

The above sections have discussed some representative examples of the control circuit of brushless servo motors in a manner that can be easily understood. There are many methods for detecting the rotor position and the speed.

Compared with the dc servo motor, the control circuit of the brushless servo motor can be said to be complicated. Likewise, the brushless servo motor has

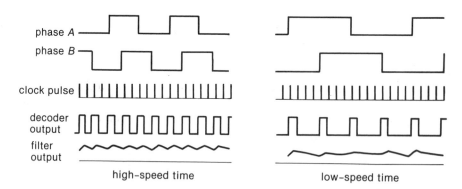

Figure 2.17 Principles of *F/V* conversion

several advantages over the dc servo motor for the user, because the former has a lower cost and a simpler structure than the latter.

Control of Brushless Servo Motors by Microcomputers (DSPs)

In the past, positioning by using a servo motor has been mainly performed by the so-called serial pulse train method. This is a method in which the numbers of command and encoder pulses are compared and the difference between the two becomes the motor speed command.

Recently, however, positioning control is performed by the CPU of microprocessors because the techniques of microcomputers have been applied to a wider range of fields than before. When the servo driver receives analog signals as inputs, the results of the CPU's operations are converted from analog to digital and used as inputs. Therefore, in the total system performing positioning control, CPU can directly control the part where the servo driver receives analog signals as inputs.

Consequently, the following merits can be obtained:

1. As all the inputs are digitized, the positioning control is not influenced by the temperature; thus, it is highly reliable.
2. It is possible to change the parameters automatically according to the load conditions. This is impossible when the inputs are analog.
3. By controlling the whole system by one CPU, the cost and unit size can be reduced.

In the following sections, the hardware constituting the control circuit of the brushless servo motor and the algorithm of the software running on the hardware, in cases where total digital control is the target, are examined for each circuit controlled by means of analog inputs.

Rotor Position Detection and Sine Wave Generation Circuit

This circuit is traditionally controlled by means of mostly digital inputs. This method was also described in the explanation of the principles of brushless servo motors. It consists of reading a sine wave table from ROM, with the signals from the 8-bit absolute encoder as addresses.

In situations where the same operation is performed by the CPU, the signals from the encoder are converted into pure binary numbers inside the CPU, regardless of what code system the signals have, and the hardware only has to convert signal levels.

In order to take advantage of the flexibility of using a CPU, sensors capable of judging minimum necessary rotor position should be attached and the motor should be started by using the signals from the sensors. In such a case, complicated absolute encoders or resolvers do not have to be used.

After the motor is started, the counter is cleared when the first index signal (the pulse that is fed once by the encoder per rotation) is found. Then, the incremental pulses are counted, the count values are read and made addresses of the sine wave table.

Speed Control Circuit

The servo driver, which is the heart of the control circuit of a servo motor, is called the servo amplifier because the deviation amplifying function of this circuit is responsible for the performance of the whole system. When the circuit receives analog signals as inputs, the difference between the reference speed and the speed feedback signal is amplified. The structure of the circuit is shown in Fig. 2.18.

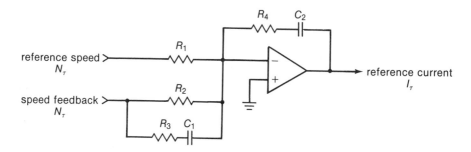

Figure 2.18   PID amplifier

This is a rather simple circuit, essentially consisting of one operational amplifier. The ratio of R_1 to R_4 represents the proportional gain, and R_4 and C_C decide the differential time constant. This is the so-called PID controller.

This can be expressed by the following equation:

$$Ir = Kp(N_r - N_f) + Ki \int (N_r - N_f)\, dt + K_D d(N_r - N_f)/dt \qquad (2.1)$$

where

$$Kp, Ki, K_D = \text{the constants of the PID.}$$

It is not appropriate for the CPU to perform this calculation as it is from the aspect of computing time. Therefore, the error is calculated once and the result is stored in memory. Then it is calculated with the result of calculation performed the next time.

In other words, Eq. (2.1) can be transformed into the following equation, where $\Delta N_n = N_r - N_f$, the time one interval before is ΔN_{n-1} and the time two intervals before is N_{n-2}.

$$I_{rn} = K_k(\Delta N_n - \Delta N_{n-1}) + K_i(\Delta N_n)$$

$$+ K_D(\Delta N_{n-2} {}^*\Delta N_{n-1} + \Delta N_{n-2}) + I_{rn-1} \qquad (2.2)$$

As this shows, the time interval required by the CPU to perform the same calculations in regular periods is called the sampling time. It is an important factor for selecting the CPU and in deciding its requirements in relation to the response time of the servo motor. In general, the sampling time should be $\frac{1}{10}$ of, or below the step response.

Current Control Circuit

Since the result of operation of the speed control part becomes the direct current reference, it is multiplied by the amplitude constant drawn from the sine wave table. The current reference of the sine wave can be obtained by this operation. The current feedback signals are detected in the analog form, so two A/D converters with a capacity of approximately 8 bits are used to read the signals by the CPU.

Since the circuit structure of the analog current controller is of PI control, as in the case of the speed control, its calculation algorithm can be considered in the same manner as that of the speed control. The current response, however, is several times faster than the speed response, so other routines having shorter sampling times are required for three-phase operation.

The brushless servo motor, like the dc servo motor, is also influenced by armature reaction due to load currents. This is a phenomenon in which the magnetic flux induced by the increase of current flowing in the coil bends the direction of the main magnetic flux by interfering with the main magnetic flux of the field. That disturbs the orthogonal relationship between the current and the magnetic flux and decreases effective torque.

If it is possible to change the phase that reads sine waves from ROM by the current value detected by the CPU, the influence of armature reaction can be suppressed.

PWM Circuit

This circuit converts the output of the current amplifier into a pulse width proportional to the output. That part feeds the result of calculations performed in the CPU outside. When the circuit receives analog signals as inputs, the outputs of the current amplifier and the chopping waves are compared using a comparator.

In cases where this is performed by CPU, it is not necessarily impossible to create the pattern of carrier waves inside the CPU and to perform the comparison operation. However, as there is a need to reduce the burden on the CPU, the following method, a combination of software and hardware, is desirable:

1. Connect a counter to the CPU and load the CPU with the result of the operation.
2. At the same time that the loading takes place, an external flip-flop is set.
3. The counter is decremented by one with each of the pulses of a given period.
4. When the contents of the counter become zero, the flip-flop is reset.

As a result of following the above procedure, the output of the flip-flop has a pulse width proportional to the result of the operation, so it follows that PWM has been performed. In addition, as the CPU only sends data to the counter, its work load is not extremely severe even when it controls three phases.

Speed Detector

In order for the CPU to calculate the speed from the pulses produced by the encoder, the frequency of the pulses should be converted to digital values. In general, the speed can be estimated by triggering the counter with pulses and reading the contents of the counter at regular intervals; however, in order to have a wide range of speed control ratio (which is a required feature of the servo motor), the count value per period has to change drastically in accordance with the speed.

Now, in this example, the recognizable rotational frequency range will be estimated when the sampling period is 2ms and an encoder developing 2000 pulses per one rotation is used.

At least ten encoder pulses are necessary in the sampling period. In this case, there will be errors within 10 percent, depending on the timing of reading even if rotational speed does not change.

In cases where the encoder has a two-phase output with a phase difference of 90 degrees, it is possible to make the frequency four times higher, so in one rotation it is possible to output 8000 pulses.

The recognizable minimum rotational frequency can be calculated by the following equation, as it equals the rotational frequency when ten pulses are counted in 2 ms:

$$\frac{\dfrac{\text{content of counter}}{\text{sampling period}} \times 60}{\text{\# of pulses from encoder}} = \frac{\dfrac{10}{0.002} \times 60}{8000} = 37.5 \text{ rpm} \qquad (2.3)$$

In other words, as long as the error is kept within 10 percent under this condition, the minimum rotational frequency is 37.5 rpm. Even a motor having a rated speed of 3750 rpm can obtain a speed ratio of only 100:1.

Needless to say, the minimum rotational frequency decreases if the sampling period is made longer or if the number of pulses of the encoder is increased. The sampling period of 2 ms is nearly the maximum of a servo motor, and the response frequency of the 2000-pulse encoder is almost the limit at the maximum rotational speed.

Accordingly, a different method should be used for a high-performance servo motor that has a speed variation ratio exceeding 100:1. In the method described in this section, the number of pulses during a given period is counted as a means of obtaining information about the speed from encoder pulses.

On the other hand, there is a method that measures the period of one pulse. The clock pulse during a given period is counted by using the reading edge of the pulse as a reference. The counting is completed at the next reading edge. In this method, the count value is in inverse proportion to the speed, and the lower the rotational speed is, the more accurately the detection is performed.

Therefore, it is necessary for a servo motor which requires a wide speed control range to use the above two methods continuously in a proper way. It can be performed as follows by utilizing the operational functions of the CPU (this is explained in Fig. 2.19): The encoder and the clock pulse are counted by counters separately with the leading edge of the encoder pulse being the starting point. The counting is stopped at the first leading of the encoder pulse after a given period T. In this case, the speed N can be calculated from the following equation:

$$N = K(P_E/P_C) \qquad (2.4)$$

where the number of encoder pulses is P_E and that of clock pulses is P_C and K is a constant.

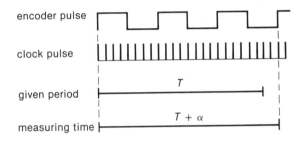

Figure 2.19   Principle of speed detection

In the above sections, some important points about controlling a brushless servo motor by microcomputers have been described. However, the focus of the explanation has mainly been on the hardware.

The object of control by microcomputers is to introduce modern control theory. This theory can be realized by the fact that the brushless servo motor is called the software servo motor. As the throughput of the CPU has been improved year after year, advanced algorithm will be applied to the control of brushless servo motors.

2.2 SENSORS AND DSP SENSORS*

The sensor attached to a brushless servo motor, in combination with an appropriate control circuit, detects the position information in positioning control. It is not until the sensor is combined with the control circuit that it becomes possible to detect all of them. This section presents features and usages of rotary encoders and resolvers.

* See references [2.3], [2.4], and [2.5].

Optical Encoders

Optical encoders are classified into the incremental type and absolute type by function. An optical encoder basically consists of a light source for emitting light, a light receiving device, and rotary disc with slits. The optical encoder obtains the pulse output proportional to the angle of rotation by turning the rotary disc. The rotary disc is placed between the light source and the light receiving device.

An example of optical encoders is illustrated in Fig. 2.20. It employs a light emitting diode (LED) as a light source. It is equipped with a fixed board with slits to guide the light through the rotary disc correctly to the light receiving device.

The light receiving device detects the light (infrared rays) projected by the LED after the light has passed through the slits of the rotary disc and those of the fixed disc. The output of the LED, which is used as the light source, varies greatly according to the temperature.

The relation between the ambient temperature and the relative light output is shown in Fig. 2.21. As the ambient temperature rises, the relative light output declines. The most common method to compensate the decline is to place two light receiving devices on the same truck and make them reduce signal output having an electrical phase difference of 180 degrees. The signal is processed by the comparator to obtain stable rectangular wave output.

The wavelength characteristics of light emitting devices, light receiving devices, and other related devices used for the optical encoder are shown in Fig. 2.22.

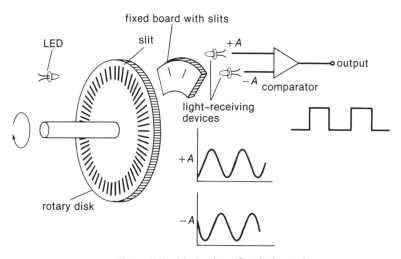

Figure 2.20 Mechanism of optical encoders

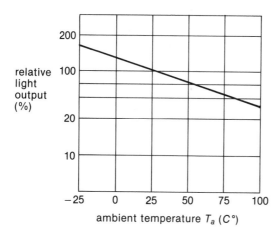

Figure 2.21 LED temperature characteristic curve

Incremental Encoder

Mechanism of incremental encoders.   Figure 2.23 is a structural drawing of an incremental encoder's relative angular position detector of output A and B type, having signal Z of phase zero.

The light projected from the LED goes through the slits of the rotary disc and each of the slits A, B, and Z of the fixed board with slits. Then the light is

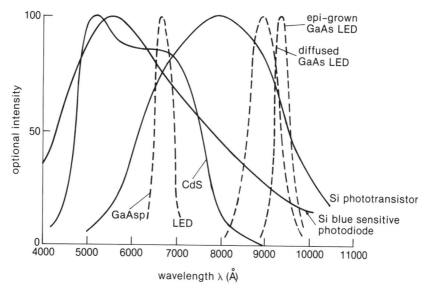

Figure 2.22 Wavelength characteristics of light-emitting and receiving devices

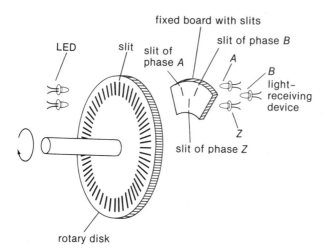

Figure 2.23   Structure of incremental encoders

detected by the light receiving devices *A*, *B*, and *Z*. Slits *A* and *B* on the fixed board with slits have a phase difference of 90 degrees. The electric output, with its waveform shaped, is a rectangular wave with the same 90-degree phase difference.

Figure 2.24 shows the condition of the final output corresponding to each of the light receiving devices. Figure 2.25 and Fig. 2.26 illustrate the principles of determining the direction of rotation.

The incremental encoder has a simple structure and is relatively inexpensive. Besides, it can easily transmit signals because it has fewer electrical wires for output. The output pulses of the encoder do not show the absolute value of the rotational position of the axis. The number of pulses are proportional to the rotational angle of the axis. The absolute value of the rotational positon of the axis is shown by the result of accumulating the output pulses of the encoder by using a counter.

The following points should be noted in using an incremental encoder. First, measures should be taken to reduce noises because noises are accumulated by the counter while the signals are transmitted. Second, the power source should

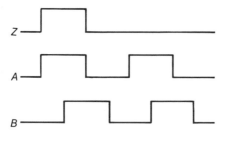

Figure 2.24   Output waveform of incremental encoders

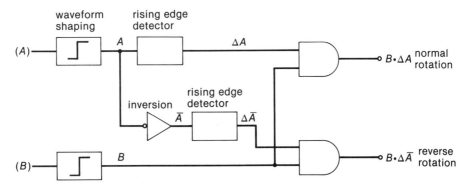

Figure 2.25 Block diagram of the determination of the direction of rotation

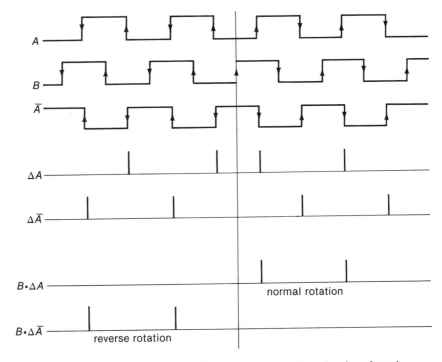

Figure 2.26 Waveform diagram of the determination of the direction of rotation

not be disconnected, for it is impossible to show the previous position once the power source is disconnected, even if it is immediately reconnected.

Detection of Rotational Speed by Incremental Encoders

The incremental encoder does nothing but generate pulse trains. Therefore, the number of output pulses of the encoder should be converted to analog signals proportional to pulse frequency by means of a F/V converter, as shown in Fig. 2.27. That makes it possible to obtain analog signals for detecting the rotational speed.

In cases where the number of pulses are too small to cause some difficulties in practical use, the number of pulses are increased by means of frequency multiplication by four (as shown in Fig. 2.28) and then converted to analog signals by a F/V converter. Figure 2.28 is an example of a frequency multiplication by four circuit.

Absolute Encoder

Mechanism of absolute encoders.   The basic structure of absolute angular position detector encoders is the same as that of incremental encoders. The pattern of the slits of the rotational disc is shown in Fig. 2.29. Slits for the necessary number of bits are arranged in a concentric circle toward the center. In that case, the most outer circle of the rotational disc shows the smallest bit.

Figure 2.30 shows the structural drawing of an absolute encoder. As the name tells, the absolute encoder is able to detect the absolute position of the input axis. Therefore, it does not accumulate errors because of the noises generated when signals are transmitted. In addition to that, it does not miss the previous position even if the power source is disconnected, while the incremental encoder does. Thus, it always detects the correct value at the time.

The defect of the incremental encoder lies in the difficulty of making it small-sized and producing it at low cost. That is because the number of output signal wires increases as the number of bits increases.

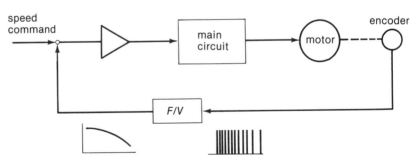

Figure 2.27 Speed detector

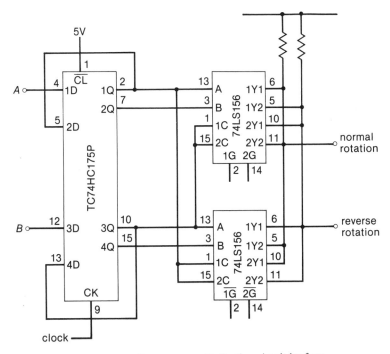

Figure 2.28   Frequency multiplication circuit by four

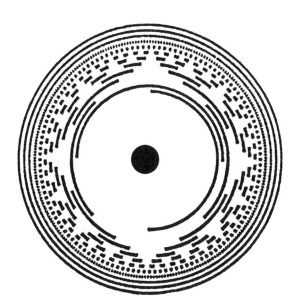

Figure 2.29   Rotary disk of absolute encoders

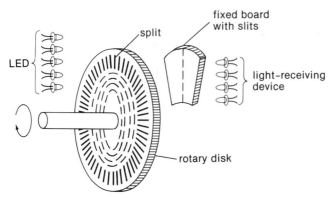

Figure 2.30 Structural drawing of absolute encoders

Output Codes of Absolute Encoders

Output codes of absolute encoders are roughly classified into binary codes and BCD codes. Binary codes are divided into natural binary codes and cyclic (Gray) binary codes.

Table 2.1 shows the difference between natural binary and cyclic binary. As shown in the table, cyclic binary is a code system in which only one bit of code changes in any case when one number changes to the next number.

Natural binary follows some ambiguity in reading the numbers when one number is changing to the next number. Cyclic binary is a code system invented to eliminate such ambiguity.

Figure 2.31 shows an example of circuits that convert cyclic binary codes into natural binary codes.

Magnetic Encoder

The optical encoder generates pulses by rotating the rotary disc between the light emitting device and the light receiving device. The magnetic encoder consists of a magnetized drum and a magnetic resistor placed close to the magnetic drum.

As shown in Fig. 2.32, the magnetic drum and the magnetic resistors are placed in two different ways. One way is to magnetize the outer circle of the drum and put the magnetic resistor so as to face that of the outer circle. Another way is to magnetize the side of the magnetic drum and put the magnetic resistor so as to face that side. Both of these methods have the same basic principle.

As described above, the magnetic encoder and the optical encoder have the same principle for the production of output signal, though they have different structures for detectors.

TABLE 2.1 BINARY CODES

Natural binary						Cyclic binary						Decimal
0	0	0	0	0	0	0	0	0	0	0	0	00
0	0	0	0	0	1	0	0	0	0	0	1	01
0	0	0	0	1	0	0	0	0	0	1	1	02
0	0	0	0	1	1	0	0	0	0	1	0	03
0	0	0	1	0	0	0	0	0	1	1	0	04
0	0	0	1	0	1	0	0	0	1	1	1	05
0	0	0	1	1	0	0	0	0	1	0	1	06
0	0	0	1	1	1	0	0	0	1	0	0	07
0	0	1	0	0	0	0	0	1	1	0	0	08
0	0	1	0	0	1	0	0	1	1	0	1	09
0	0	1	0	1	0	0	0	1	1	1	1	10
0	0	1	0	1	1	0	0	1	1	1	0	11
0	0	1	1	0	0	0	0	1	0	1	0	12
0	0	1	1	0	1	0	0	1	0	1	1	13
0	0	1	1	1	0	0	0	1	0	0	1	14
0	0	1	1	1	1	0	0	1	0	0	0	15
0	1	0	0	0	0	0	1	1	0	0	0	16
0	1	0	0	0	1	0	1	1	0	0	1	17
0	1	0	0	1	0	0	1	1	0	1	1	18
0	1	0	0	1	1	0	1	1	0	1	0	19
0	1	0	1	0	0	0	1	1	1	1	0	20
0	1	0	1	0	1	0	1	1	1	1	1	21
0	1	0	1	1	0	0	1	1	1	0	1	22
0	1	0	1	1	1	0	1	1	1	0	0	23
0	1	1	0	0	0	0	1	0	1	0	0	24
0	1	1	0	0	1	0	1	0	1	0	1	25
⋮	⋮	⋮	⋮	⋮	⋮	⋮	⋮	⋮	⋮	⋮	⋮	⋮
⋮	⋮	⋮	⋮	⋮	⋮	⋮	⋮	⋮	⋮	⋮	⋮	⋮
⋮	⋮	⋮	⋮	⋮	⋮	⋮	⋮	⋮	⋮	⋮	⋮	⋮
⋮	⋮	⋮	⋮	⋮	⋮	⋮	⋮	⋮	⋮	⋮	⋮	⋮
⋮	⋮	⋮	⋮	⋮	⋮	⋮	⋮	⋮	⋮	⋮	⋮	⋮
⋮	⋮	⋮	⋮	⋮	⋮	⋮	⋮	⋮	⋮	⋮	⋮	⋮
1	1	1	1	1	1	1	0	0	0	0	0	63
32	16	8	4	2	1			—				—

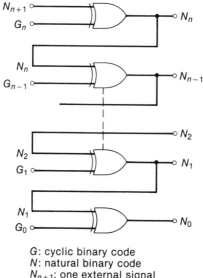

G: cyclic binary code
N: natural binary code
N_{n+1}: one external signal
        input terminal

Figure 2.31   Code converter

Operational Principles

The magnetic resistor is a magnetism-electricity converting device whose electric resistance varies according to the intensity of the applied magnetic field (see Fig. 2.33). The magnetic resistor has resistance R when current i flows in it in the direction of the arrow. R goes down when magnetic field H is applied at a right angle to the direction of the current i. Accordingly, it is possible to obtain signal output corresponding to the rotation of the drum by placing the magnetic resistor so that it corresponds to the magnetizing pattern of the drum.

    The magnetic resistor is sensitive to temperature; therefore, the use of only one magnetic resistor tends to cause drift of the output. The varying condition of

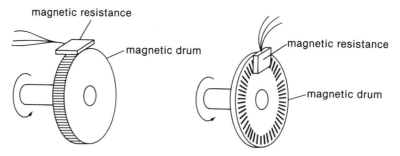

Figure 2.32   Structural drawing of magnetic encoders

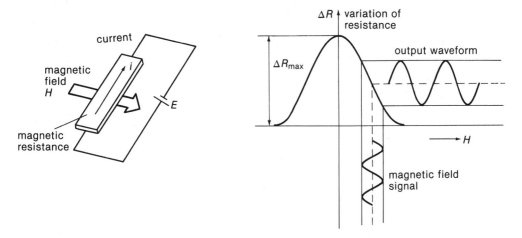

Figure 2.33   Variation of electric resistance and output waveform

direct current resistance due to the change of the surrounding temperature is shown in Fig. 2.34.

The variation of the output due to temperature change has to be reduced as much as possible. For that purpose, magnetic resistors are combined to make a bridge circuit, which improves the temperature characteristics, and simultaneously, output voltage is increased.

In the case of using the magnetic encoder as an incremental encoder, it is equipped with another circuit of the same structure so that the two circuits have a phase difference of 90 degrees in electrical angle. That enables the encoder to obtain the output of phases A and B. If necessary, the output of phase Z can also be obtained by applying that method.

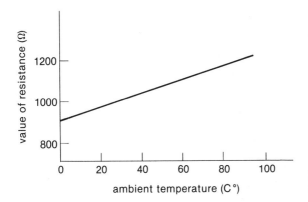

Figure 2.34   Variation of magnetic resistance due to temperature change

Features of Magnetic Encoders

In using encoders, we need to know about their characteristics. Sensitivity to environment, resolution, responsibility, and so on, should be the criteria of selection. Magnetic encoders have the following characteristics:

1. Insensitive to environment, influenced less by dust, dew condensation, and so on.
2. Have high resolution (1000–3000 P/rev is the standard).
3. Have high responsibility (no decline of output up to around 200 kHz).
4. Are simple in structure.

Cautions for Using Magnetic Encoders

Magnetic encoders are operated by applying magnetic intensity. In using them at places with intense magnetic field or much magnetic powder, the following points should be taken into consideration:

1. They malfunction when intense magnetic field is applied to them. In such case, their magnetic intensity should be protected by a shielding case made of magnetic substance.
2. They malfunction when magnetic powder enters the drum and sticks to it. When they are used in such places, they should be protected by a case which prevents the entrance of magnetic powder.

Cautions for Using Rotary Encoders

In choosing rotary encoders, the purposes of application should be specified. Those that answer the purposes should be selected. The points to be checked in choosing them are as follows:

1. resolution . . . for precision of position and speed control ratio
2. size . . . for condition of attachment
3. allowable load on the axis . . . for life and condition of attachment
4. maximum allowable rotational frequency . . . maximum response frequency
5. phase difference of output
6. sensitivity to environment . . . for place of attachment and so on
7. incremental type or absolute type?

Varieties of Output Circuits

1. **TTL level.** When the level of output is for TTL, the source voltage of the IC is low (generally 5 V), so the output is easily affected by noises. Therefore, it is disadvantageous to use the circuit in places surrounded by strong noise. In

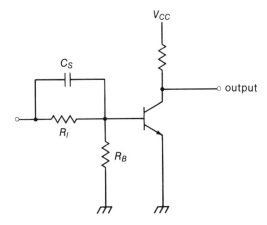

Figure 2.35 Circuit for shortening the rise time of pulses

order to prevent the output from being affected by noises, sufficient protection of adding a line driver should be made.

2.   Output of the open collector circuit. When the open collector circuit is applied, it is free to choose source voltage to a certain degree. Accordingly, the level of transmission should be decided in consideration of the surrounding noise level. The rise time of the pulses becomes longer when the distance of transmission is made longer. In order to prevent that, special attention should be paid to rise time vs transmission distances. An example of measures to shortern the rise time of pulses is shown in Fig. 2.35.

Wiring of Rotary Encoders

Electrial wires for signal transmission should be selected in consideration of the surrounding noise level. In general, shielded cable (Fig. 2.36), twisted pair shielded cable, or the equivalent are used.

Cautions for Coupling

In coupling a rotary encoder with a motor, the shafts of both the encoder and the motor should not be eccentric. Eccentric shafts degrade the encoder's precision as well as shorten the mechanical life of same.

The shapes of encoders are classified into two types, shown in Fig. 2.37 and Fig. 2.38. The shape shown in Fig. 2.38 is called the hollow shaft type.

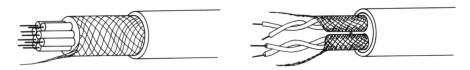

Figure 2.36   Shielded cable

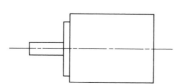

Figure 2.37 Encoder of shaft type

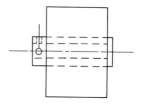

Figure 2.38 Encoder of hollow type

The method of coupling differs according to the shape of encoders. The methods illustrated in Fig. 2.39 and Fig. 2.40 are the typical methods. The method in Fig. 2.39 tends to cause insufficient clamping of set screws. In case the encoder to be coupled is of the hollow shaft type, a bellows coupling cannot be used because the shaft of the encoder is connected directly to the shaft of the motor. In that instance, the case of the encoder and the bracket of the motor are connected by a thin leaf spring.

Resolver

A resolver is a detector of the position and the angle of rotation and is used as a sensor for the motor controller. The encoder converts the amount of displacement into the digital amount, while the resolver converts it into analog amount. In general, a resolver equipped with a rotary transformer is called the brushless resolver. The structural drawing of the brushless resolver is shown in Fig. 2.41.

The brushless resolver consists of a stator, a rotor, and a rotary transformer. The windings of the stator and the rotor are distributed so that the magnetic flux is distributed in the form of a sine wave to the angle of rotation.

The stator winding, which is the excitation winding, is of two-phase structure, with an electrical phase difference of 90 degrees. The rotor winding, which is the output winding, is either one-phase or two-phase winding, depending on the use. Figure 2.42 shows the connection diagram and the relational expression of a resolver of single-phase output.

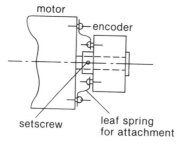

motor

encoder

setscrew   leaf spring
for attachment

Figure 2.39 Coupling by means of a bellows coupling

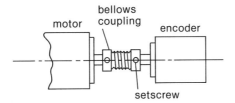

Figure 2.40   Coupling by means of a leaf spring

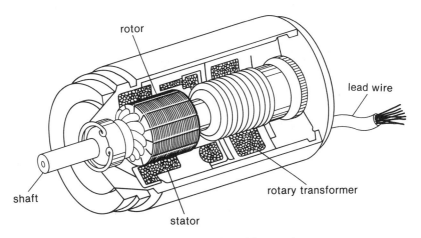

Figure 2.41   Structure of brushless resolvers

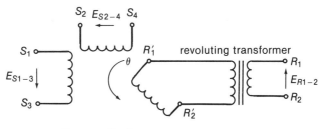

$$E_{R1-2} = K_1 (E_{S1-2} \cos \theta - E_{S2-4} \sin \theta)$$

K_1 : ratio of voltage transformation
θ : rotational angle of resolvers

Figure 2.42   Connection diagram and relational expression of output for resolvers of single-phase output

Detector

1.   Detection of position signal. Figure 2.43 shows an example of a circuit converting the signal of a resolver into digital signal. The conversion circuit in this example uses a dedicated converter, capable of exciting the stator and processing the output of the resolver. The output signal of the converter is widely used as the signal for detecting absolute values, the incremental signal, and the input signal to computers.

2.   Detection of speed signal. The methods of detecting speed by using a resolver are as follows:

One method is to count the incremental signals per unit period obtained by the circuit shown in Fig. 2.43. Another method is to calculate the difference of position signals for detecting the absolute value per unit period. The latter method is called the differential method.

Characteristics of Resolvers

The structure of resolvers is similar to that of motors. Resolvers have lower sensitivity to environment. Their characteristics are as follows:

1. Highly resistant to environmental conditions, such as vibration and impact.
2. Usable in a wide temperature range.
3. Capable of long-distance signal transmission.

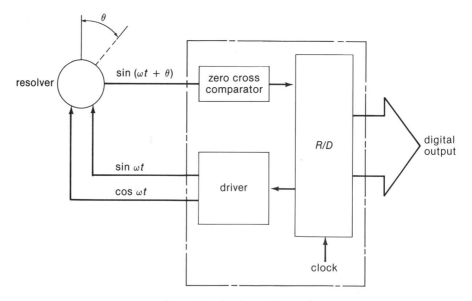

Figure 2.43   Conversion circuit applying a *R/D* converter

4. Miniaturizable.

5. Complicate the signal processing circuit.

6. Production costs are higher than for rotary encoders.

Rotary Encoder Terminology

The following terms are especially important for rotary encoders.

1. Resolution: capability to divide one rotation by n. Incremental encoders express the resolution by the number of pulses produced per one rotation. Absolute encoders express the resolution by the number of binary digits together with the number of division; 8 binary digits divided by 256, for example.

2. Maximum response frequency: the maximum frequency in which rotary encoders can electrically respond. Incremental encoders express it by the number of output pulses per second.

3. Maximum allowable rotational frequency: the maximum rotational frequency that is mechanically allowable for rotary encoders.

4. Phase A output* phase B output: the output of incremental encoders. The ideal phase difference between phase A and phase B is 90 degrees + 0 degrees. The standard of practically applied rotary encoders is with a phase difference of 90 degrees + 45 degrees.

5. Phase Z output: this is one pulse signal per revolution. It is mainly used together with an incremental type of encoder. It represents the origin.

6. Starting torque: the torque generated at the beginning of rotation.

7. Moment of inertia of the axis: the moment of inertia about the axis of the rotor. The smaller it is, the more easily and quickly the rotor is started or stopped.

8. Allowable load for the axis: the allowable load in the radial direction that is vertical to the axis (radial load) and in the thrust direction that is parallel to the axis (thrust load).

All calculations required in implementing sensors can be performed by DSPs. Therefore, a good sensor at low cost can be made. An observer described in Chapter 6 is a kind of digital filter implemented with a DSP. This may be considered as a sensor. The signal from sensor is also processed with a digital filter implemented with a DSP described in Chapter 6.

2.3 CONTROLLERS AND DSP CONTROLLERS

A digital controller is actually a signal processing system that executes algebraic algorithms inherent to the control of feedback and feedforward systems (i.e., compensator and filter algorithms).

TABLE 2.2   PRESENT AND FUTURE DSP

Signal processors		Type	Available
NEC	μPD 7720	U	1982
Texas Instr.	TMS 32010	U	1983
Fujitsu	MB 8764	U	1984
STC	DSP 128	U	1985
Texas Instr.	TMS 32020	U	1985
Texas Instr.	TMS 320C25	U	
Nat. Semi.	LM 32900	UC	
Analog Dev.	ADSP 2100	UC	
Phillips	PCB 5011	U	
Thomson	TS 68930	U	
Motorola	DSP 56000	U	
Nat. Semi.	LM 628	A	
NEC	μPD 77220	U	
NEC	μPD 77230	U F	

U universal
C processor core (external memory)
A algorithm-specific
F floating-point arithmetic

Digital single-chip signal processors (DSP) are a very attractive means for the implementation of measurement and control algorithms, mainly because of their computing speed, which is more than an order of magnitude higher than with fast modern 16/32-bit microprocessors or microcontroller using fixed-point arithmetic and whose architecture is very close to that of a personal computer. Table 2.2 and Table 2.3 show available DSPs.

Therefore, the DSP is appropriate for use in controlling fast dynamic systems, such as mechanical and complex systems (multivariable systems), and is suitable for implementing sophisticated control algorithms that have been discarded for decades due to insufficient computation speed. The recent advanced motor control technologies require faster and more accurate control performance.

TABLE 2.3   ACHIEVABLE SAMPLING FREQUENCIES

Microprocessor	Clock	fs
8086	8 MHz	<2 kHz
Z8000	5 MHz	<2 kHz
68000	10 MHz	<4 kHz
32016	10 MHz	<5 kHz
TMS32010 signal processor		31 kHz

In order to cope with this, the well-developed and extended modern control theory yields excellent control performance. Fast DSPs can be used to practically and effectively implement such theory.

The Texas Instruments TMS 320 family provides several beneficial features for implementing digital control system elements through its architecture, speed, and instruction set.

A prominent feature of the TMS 320 family is the on-chip, 16×16-bit multiplier performing two's-complement multiplication and producing a 32-bit product in a single 200-ns instruction cycle. The TMS 320 family instruction set includes special instructions necessary for fast implementation of sum-of-products computations, encountered in digital filtering/compensation and Fourier transfer calculations. Most of the instructions critical to signal processing execute in one instruction cycle.

BIBLIOGRAPHY

[2.1] Y. Dote, "Small-sized Control Motor," *Daily Kogyo Shimbun,* September 4, 1984.

[2.2] Y. Dote, "Varieties and Characteristics of Control Motors," *Machine Design,* vol. 27, no. 3, September 1983, pp. 29–34.

[2.3] Catalog from Tamagawa Seiki Co. Ltd., "Shaft Encoders and the Application of Them," 1983.

[2.4] Triceps Co. Ltd., "Optimal Usage of Encoders and Resolvers, and the Application of Them to ac Servo Motors," 1983.

[2.5] Omron Tateishi Electronics Co., Catalog of Rotary Encoders, July 1984.

Chapter 3

Control Theory Overview

3.0 INTRODUCTION

There has been a gap between modern control theory and practice. However, a practicing engineer who knows control systems well has modified, extended, and combined the modern control theory with classical control theory. Successful results using developed control theory, especially to servo motor and motion, have been reported. In this chapter, practical and useful control theories or methods are introduced. The theory (method) may be given in the S domain (continuous); however, by using the bilinear transformation described in section 7.2 (this is possible when implementing with a DSP), it can be replaced by a discrete (digital) control method. Digital controllers are actually digital filters; therefore, digital signal processing (digital filter) is also overviewed. Control theory terminologies used in this book are illustrated in order to easily follow the book.

3.1 CONTROL THEORY AND SIGNAL PROCESSING

Control Theory Overview*

"To control" is defined as "to manipulate an object (motor) so as to serve a certain purpose" (so as to make it work as required).

* See reference [3.1].

In the aforementioned case, the motor is called the controlled system. The physical quantities (voltage, current, frequency, torque, angular velocity, and angle) are called the controlled variables, the required manipulations are called the controls (voltage, current, and frequency), and the instruction to be given is called the control reference.

Automatic control means to perform the control not by hand but by equipments and machines. Automatic control has two varieties. One is sequential control (sequential starting and overload protection of motors). It applies a sequential circuit which is comprised of a memory circuit and a logical circuit. Another variety controls the dynamic system including energy storage elements (inertia, inductor, capacitor, and so on). This book will provide information about automatic control of the dynamic system.

Control is classified into feedback control and feedforward control. Feedback control detects the controlled variables, then compares them with the control reference to determine the control variables.

Feedback control is insensitive to disturbances (load torque fluctuation, source voltage fluctuation, etc.) which disturb the behavior of the system, and to the parameter variation (change in inertia, resistance, etc.). Feedback control also changes the structure of control systems.

Feedforward control makes the response of the control system faster because it determines the controls using future information. Example: When you drive a car on the road, you will see what the road ahead looks like. The control reference is called desired value or input, and the controlled variable is called output. The difference between the desired value and the controlled variable is called error.

Control has the regulator problem (constant angular velocity control), the servo problem (point to point control), and the tracking problem (continuous path control). The regulator problem is the stability of the system when the control is zero. The servo problem is the control of the system when the control varies in respect to time is considered. The tracking problem is the utilization of future information for control, when the control instruction varies with passage of time and the condition of the variation is known beforehand. A compensator is able to convert the servo problem into the regulator problem.

The value of error becomes constant as time passes. That value is called steady-state error, and the movement of the error, until it becomes constant, is called transient response. The steady-state error should be smaller (for superior precision) and the transient response should be quicker; there is a constant trade-off between the two.

The recent, rapid, and revolutionary progress in power electronics and microelectronics has made it possible to implement and apply modern control theory, well developed, modified, and extended over the last decades, to various kinds of motor controls. An overview of digital control problems arising in the usage of microprocessors (essential hardware for motor control) and the application of modern control theory (software in motor control) is given.

Modern control theory, which is suitable for attacking complex control systems, has recently been widely applied to design and analysis of controllers for motors. The reasons are as follows:

1. The necessity of meeting increasingly stringent requirements on the performance of motor control systems and their own system complexity.
2. Easy access to modern microprocessors and DSPs with which very sophisticated control strategies can be implemented at reasonable cost.
3. With the advent of the power semiconductor switching components and LSI's, it has become possible to construct a higher performance drive system with microprocessors and sensors.
4. Modern control theory has been extended and modified so that it is practically applicable to microprocessor-based motor control by taking into account physical constraints, such as input delay times and input and state variable constraints.

Microprocessors (μp) began to be applied in controllers of drive electronics systems in the late 1970s. They were first used for large power dc drives for steel mills and large power ac drives in energy saving applications.

Currently, microprocessor-based controllers are commonly used in almost every application area of drive electronics, and most customers are not satisfied by controllers without microprocessors. General purpose 8-bit microprocessors have been, to date, most commonly used for this purpose.

However, 16-bit microprocessors have begun to be widely applied in high-performance applications. Single-chip microcomputers and custom LSI's are commercially available for special application areas. Recently, digital signal processors (DSPs) have been used because of their fast computational capability and suitable architecture for digital controllers. The multiprocessor configuration is also used for sophisticated control performance. Table 3.1 shows a performance comparison of 8- and 16-bit microprocessors. Table 3.2 exhibits possible architectures.

Control performance of drive electronics systems has been much improved by the use of microprocessors. The basic configuration of microprocessor-based controllers is shown in Fig. 3.1, and the effects of DCC (Direct Digital Control) applied in drive electronics systems are listed in Fig. 3.2.

Figure 3.3 through Fig. 3.5 show typical examples of μp-based controllers applied to drive electronics systems. Figure 3.3 shows a dc servo system using a single-chip microcomputer. Figure 3.4 shows an ac drive control system with a custom LSI for PWM pattern generation. Figure 3.5 presents a multiple 16-bit processor system, for a GTO inverter-fed vector-controlled induction motor.

There are usually three control loops included in digital controllers for small motors. The innermost loop is for current control. The outermost loop is for position control, and a speed control loop is added between the two. An inner

TABLE 3.1 COMPARISON OF TWO 8-BIT MICROPROCESSORS IN PERFORMANCE WITH 16-BIT MICROPROCESSOR

	8-Bit Microprocessor (A)	16-bit Microprocessor (B)
Configuration	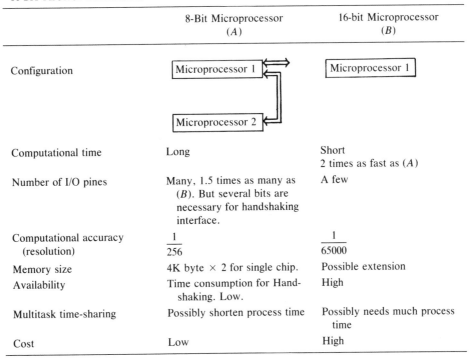	
Computational time	Long	Short 2 times as fast as (A)
Number of I/O pines	Many, 1.5 times as many as (B). But several bits are necessary for handshaking interface.	A few
Computational accuracy (resolution)	$\dfrac{1}{256}$	$\dfrac{1}{65000}$
Memory size	4K byte × 2 for single chip.	Possible extension
Availability	Time consumption for Handshaking. Low.	High
Multitask time-sharing	Possibly shorten process time	Possibly needs much process time
Cost	Low	High

loop is designed to give about five times as fast response time as the next outer loop. They roughly linearize and decouple the entire control loop.

By paying attention to the basic philosophy and classification of various types of modern control techniques, the "state of the art" and the principles of modern control theory are systematically explained.

Modern control theory is contrasted with conventional control theory, in that the former is applicable to multi-input multi-output systems, which may be linear or nonlinear and time-varying. The latter is applicable only to linear time-invariant, single-input, single-output systems. Modern control theory is essentially a time-domain approach based on the concept of state, while conventional control theory is a complex-domain approach. System design in classical control theory is based on trial-and-error procedures which, in general, will not yield optimal control systems. System design in modern control theory, however, enables the engineer to systematically design optimal control systems with respect to given performance indexes.

In addition, design in modern control theory can be carried out for a class of inputs, instead of a specific input function, such as the impulse function, step function, or sinusoidal function. Modern control theory enables the engineer to

TABLE 3.2 ARCHITECTURE OF MICROPROCESSOR

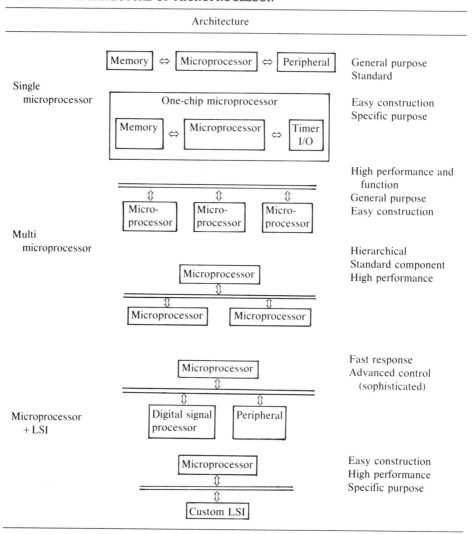

Architecture

Single microprocessor

Memory ⇔ Microprocessor ⇔ Peripheral — General purpose / Standard

One-chip microprocessor: Memory ⇔ Microprocessor ⇔ Timer I/O — Easy construction / Specific purpose

Multi microprocessor

Micro-processor / Micro-processor / Micro-processor — High performance and function / General purpose / Easy construction

Microprocessor; Microprocessor — Microprocessor — Hierarchical / Standard component / High performance

Microprocessor + LSI

Microprocessor — Digital signal processor — Peripheral — Fast response / Advanced control (sophisticated)

Microprocessor — Custom LSI — Easy construction / High performance / Specific purpose

include initial conditions in the design. The most recent developments in modern control theory may be said to be in the direction of the optimal control of both deterministic and stochastic systems as well as the adaptive, learning, and intelligent control of complex systems. Applications of modern control theory are now under way, and interesting and significant results can be expected.

Since modern control theory is quite complicated, it is considered here from four different viewpoints. They are: the control objective, design procedure, physical systems, and functions assigned to w, x, y, and z axes, respectively. Fur-

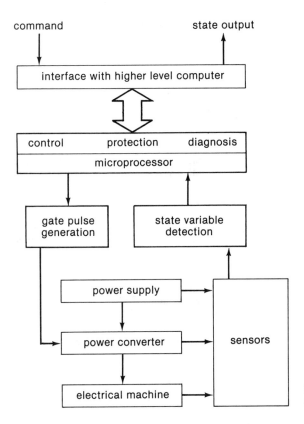

Figure 3.1 Basic configuration of microprocessor-based controller

thermore, the structure of modern control theory is explained by using a coordinate system and some set (w, x, y, z) in the four-dimensional space, as shown in Fig. 3.6. The surveyed papers are classified into six categories shown and underlined in Fig. 3.6. They are optimal control, robust control, feedback-feedforward control, passive adaptive control, active adaptive control, learning and intelligent control, and distributed parameter control.

The control objective (w axis) is to determine inputs to a physical process in order to achieve desired goals, such as obtaining some optimal controller in spite of parameter variations and disturbances which are present (robustness). A large number of the surveyed papers that deal with optimal and adaptive controls are available.

In solving problems of optimal control systems, one may have the goal of finding a rule for making the present control decision, subject to certain constraints which will somewhat minimize deviation from ideal behavior. In the basic optimal-control problem, a given control system is described as

$$x(n + 1) = f(x(n), u(n), n) \tag{3.1}$$

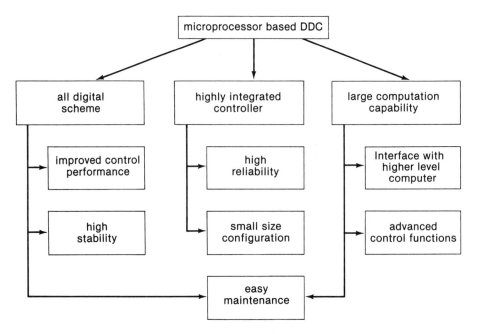

Figure 3.2 Effects of Direct digital control (DDC) applied in drive electronics systems

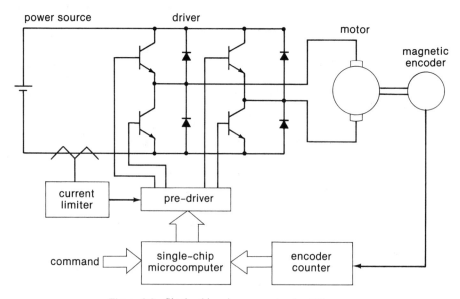

Figure 3.3 Single-chip microcomputer for DC servo

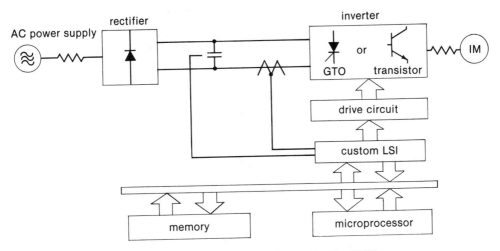

Figure 3.4   Custom LSI for PWM

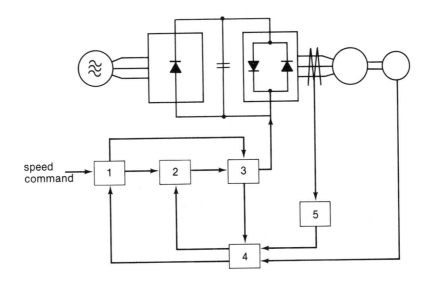

1 speed control
  slip freq detection
2 current control
  vector calculation
3 PWM generation
4 current component detection
  speed detection
5 A–D converter

Figure 3.5   Microcomputerized controller for high-performance AC drive

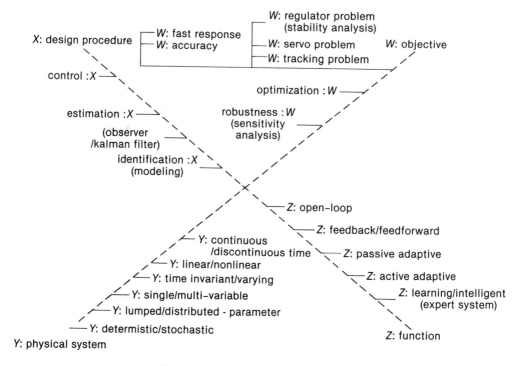

Figure 3.6 Stucture of modern control theory

The performance of this system is judged by means of an integral performance index of the form

$$PI = L(x(n), u(n), n) \tag{3.2}$$

The problem is then to find the optimal control $u(n)$ which transfers the system in Eq. (3.1) from some given initial condition or state $x = x(0)$ to some terminal condition, or final state $x = x(N)$, while minimizing the performance index in (Eq. 3.2) subject to the constraints:

$$g[x(n0] = 0, h[u(n)] = 0 \tag{3.3}$$

Most of the surveyed papers do not take system parameter variations (robustness) into account. However, an improved optimal regulator control system synthesis method has been developed to obtain a robust controller and applied to speed control for a static Scherbius induction motor system. Several physical constraints, such as input variable constraints, state variable constraints, and input delay times, have been treated simultaneously.

The robust control strategy for linear systems consists of a tracking-error driven internal model (compensator) and a stabilizer for the composite system.

BOOK MARK

Let's put New Hampshire to work, for a change!

CAROL CARSTARPHEN
DEMOCRAT, N.H. SENATE DIST. 15

CONCORD • HOPKINTON • PEMBROKE

P.O.Box 1475, Concord, NH 03302 Phone 224-7076

Carol Carstarphen
fiscal agent

The internal model is a linear time-invariant dynamic system, whose eigenvalues are those of the exogenous signals, namely the reference input and the disturbance signals. In the steady state, the tracking error is zero and the internal model generates a signal with modes identical to those of the disturbance and reference signals. A feedforward controller has the advantage that the response of the controlled system is made relatively fast, e.g., if a disturbance occurs in the system, corrective action immediately takes place. A robust controller is shown in Fig. 3.7.

A robust controller is designed so that the actual transfer function between the external disturbance and the output may match the desired one. Then, a reference input observer is determined so that the actual transfer function between the reference input and the output may also be close to the desired response. They can be designed independently; therefore, this controller is called in general a two-degrees-of-freedom linear compensator described in section 3.2.

In this control design problem, it has been assumed that a mathematical model is not available. Some simple "off-line" experiments have been applied to the system and a controller has been obtained using the results of those experiments and by applying some simple "on-line" tuning experiments.

Another robust control is obtained by utilizing zeroing and model matching

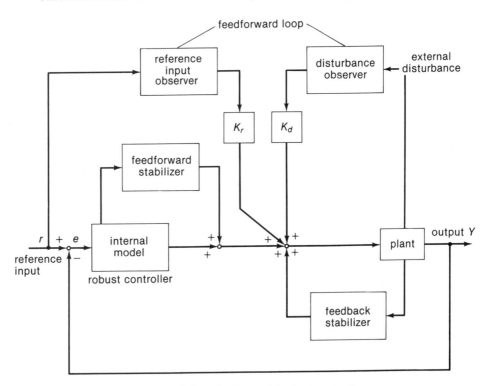

Figure 3.7   Robust feedforward-feedback controller

techniques described in Chapter 5. First, an equivalent disturbance observer is designed as follows:

The motor equation is written as

$$J_a\dot\omega + T_1 = K_t i_a$$

where J is an inertia, T_1 is an external torque, K_t is the torque constant, ω is a motor speed, and i_a is armature current.

Then, taking Laplace Transformation on the both sides of the equation and defining $T_e(s)$ to be an equivalent disturbance:

$$T_e(s) = T_\ell(s) + \Delta Js\omega_{(s)} - \Delta K_t i_a(s)$$

Then

$$s\hat{J}\omega(s) + T_e(s) = \hat{K}_t i_a(s)$$

or

$$T_e(s) = -S\hat{J}\omega(s) + \hat{K}_t i_a(s)$$

where Δ represents the nominal value and Δ is deviation of the value, putting a low pass filter $1/(s+kf)$, the estimate of $T_e(s)$, $\hat{T}_e(s)$ is obtained as follows:

$$\hat{T}e(s) = [-s\hat{J}\omega(s) + \hat{k}_t i_a(s)]/(s+kf)$$

This estimator is called an equivalent disturbance observer described in Chapter 6.

$$\frac{\hat{T}_e(s)}{\hat{K}_t}$$

is fed into the input to cancel the equivalent disturbance, including the disturbance due to system parameter variations. Next, by using the nominal parameter values, a feedback controller for model matching is designed. This zeroing and model matching technique can be explained in terms of sliding mode control with model, feedback control for model matching, and zeroing corresponding to sliding curve, equivalent control, and a cancellation of disturbances with an observer, respectively.

Modern control theory is essentially based on state feedback control, which changes system structures and makes systems insensitive to system parameter variations and disturbances. Feedforward control is often applied to preview control (tracking problem) and decoupled control.

Interest in adaptive methods on axis z (Fig. 3.6) has recently increased rapidly, along with interest and progress in control topics in general. The term adaptive system has a variety of specific meanings, but it usually implies that the system is capable of accommodating unpredictable environmental changes, whether these changes arise within the system or external to it. This concept has a great deal of appeal to the systems designer since a highly adaptive system, besides accommodating environmental changes, would also accommodate engineering design errors or uncertainties and would compensate for the failure of minor system components, thereby increasing system reliability. Because engineers are now interested in constructing robust controllers, there is a rigorous development of this field, both in universities and industries.

There are three schemes for parameter adaptive control: gain scheduling, model reference control, and self-tuning regulators. It is sometimes possible to find auxiliary process variables that correlate well with the changes in process dynamics. It is then possible to eliminate the influences of parameter variations by changing the parameters of the regulator as functions of the auxiliary variables. See Fig. 3.8(a). This approach is called gain scheduling because the system was originally used to accommodate changes in process gain only.

Gain scheduling is an open loop scheme comparable to feedforward compensation. There is no feedback to compensate for an incorrect schedule, and it has the advantage that the parameters can be changed quickly in response to process changes. This is the predominant technique for design of high-performance, industrial control systems and is in frequent, practical use.

Another way to adjust the parameters of the regulator is illustrated in Fig. 3.8(b). This scheme was originally developed for the servo problem. The specifications are given in terms of a reference model, which dictates how the process output should ideally respond to the command signal. Note that the reference model is part of the control system. The regulator can be considered as consisting of two loops. The inner loop is an ordinary control loop composed of the process and the regulator. The parameters of the regulator are adjusted by the outer loop in a manner that the error e between the model output y and the process output$^{\hat{y}}$ and the process output^y becomes small. The outer loop is similar to regulator loop. The key problem is to determine the adjustment mechanism.

Although an adaptive method deals with the control of dynamic systems, consideration of the problem of the adaptively controlling nondynamic system shown in Fig. 3.9 is helpful in understanding the algorithms used for gain adjustment in adaptive methods. In the figure, $k(n) = k_c + k_v(n)$, where k_c is an unknown constant, k_v is an adjustable gain, and $y(n) = k(n)u(n)$. The objective is to adjust $k_v(n)$ in a way that will make $y(n)$ approach zero. It is assumed that $u(n)$ is an arbitrary, uniformly bounded function. Since $u(n)$ is bounded, $y(n) \to 0$ if $k(n) \to 0$. The desired objective is achieved by adjusting $k_v(n)$ according to the algorithm $k(n + 1) = k(n) - Au(n)y(n)$ (A is a possible constant). This adjustment algorithm causes the time difference of the Liapunov function $V(n) = k^2(n)$ to be $V(n + 1) - V(n) < 0$ for $2 > u^2(n)A$; thus, $K(n) \to 0$ (hence $y(n) \to 0$) as $n \to \infty$ provided $u(n)$ is "sufficiently exciting." All of the adaptive techniques employ gain adjustment algorithms of this type, although some are modified to include a so-called "proportional term."

In general, an adaptive mechanism takes the form of new parameter = old parameter + (Bounded Step Size) × (Function of Input) × (Function of Error).

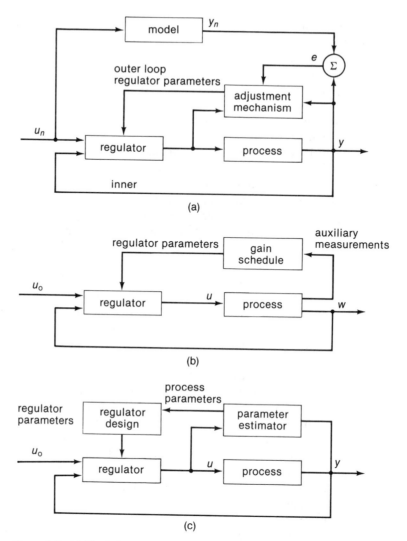

Figure 3.8 (a) Block diagram of system with gain scheduling. (b) Block diagram of model reference adaptive system (MRAS). (c) Block diagram of self-tuning regulator (STR).

Figure 3.9 Nondynamic adaptive system

However, there are the following shortcomings of MRAC, especially for fast dynamic (i.e., drive) systems:

1. Sensitivity of disturbances
2. Stability problems (richness condition, dead zone)
3. Slow time response due to mismatched initial values of integrators in the adaptive mechanism

A third way of adjusting the parameters is to use the self-tuning regulator. Such a system is shown in Fig. 3.8(c). The regulator can be considered as having two loops. The inner loop consists of the process and an ordinary linear feedback regulator. The parameters of the regulator are adjusted by the outer loop, which is composed of a recursive parameter estimator and a design calculation.

Since the input signal to the process is generated by feedback in the system [shown in Fig. 3.8(c)], there is no guarantee that good parameter estimates will be obtained. A necessary condition for parameter identifiability is that the input signal must be of the appropriate order and persistently exciting. To ensure this, it may be necessary to add excitation signals or to update the estimates only when the input is persistently exciting.

The box labeled "regulator design" in Fig. 3.8(c) represents an online solution to a design problem for a system with known parameters. This problem is called the underlying design problem. Such a problem can be associated with most adaptive control schemes. The problem is often given indirectly. To evaluate adaptive control schemes it is useful to find the underlying design problem because it will give the characteristics of the system under ideal conditions.

The self-tuning regulator is very flexible with respect to the design method, and virtually any design technique can be accommodated. To date, self-tuners based on phase and amplitude margins, pole-placement, minimum variance control, and linear quadratic gaussian control have been considered. Many different parameter estimation schemes may be used, for example: stochastic approximation, least squares, extended and generalized least squares, instrumental variables, extended Kalman filtering, and maximum likelihood method.

When checking the surveyed papers, it is apparent that a considerable gap exists between active adaptive theory and adaptive practice. However, continued interaction between theorists and practicing engineers is leading to new directions of research in the field and helping to close this gap. One solution to this problem, the author would suggest, is to use passive adaptive control, which is described in the next section.

Variable Structure Control with Sliding Mode

In many control applications, it has always been a challenge for the control engineers to design systems that will be insensitive to parameter variations and disturbances. A method called sliding mode control is suggested to overcome these difficulties (the original ideal was devised by Prof. Utkin). In sliding mode control, the representative point of the system is constrained to move along a predetermined hyperplane or hyperplanes in state space. This way, parameter insensitivity and disturbance rejection can be ensured. In order to achieve such a sliding regime, the control law is required to have a discontinuous nature, i.e., the system structure needs to be changed in time. Such a system is called a variable structure system (VSS).

The theory of VSS has been studied in great detail in Soviet literature, where it has been used to stabilize a class of nonlinear systems. However, only a few practical results have been reported. Others have been verified only by simulations. As always, a gap exists between theory and practice, requiring more detailed practical investigations to demonstrate the effects of the various terms in the control law and to set rules of thumb for design.

The shortcomings of the methods with sliding modes developed so far in literature are as follows:

A. There is a "reaching" phase in which the trajectories starting from a given initial state, away from the sliding curve, tend toward the sliding curve. Thus, the trajectories in this phase are sensitive to parameter variations and disturbances.

B. Due to switching delays, neglected small time constants, etc., ideal sliding does not occur. Instead, the trajectories chatter along the sliding curve. Such trajectories are termed nonideal sliding motions. The magnitude of the chattering is proportional to the time derivative of the sliding curve times the switching delay time.

The drawback A is removed by introducing a modified model following control with feedforward control. Since the disturbance may be considered as the time derivative of a forced model, which is to be tracked by the output variable, the time derivative of the desired output model is added to the control. This also makes a time response faster. To roughly cancel an equivalent disturbance, an equivalent disturbance observer is introduced. The feedforward gain is adjusted by applying active adaptive mechanism. Thus, this variable structure control belongs to the class of feedforward and feedback controls, which are commonly adopted for tracking problems. This also eliminates shortcoming B, since the error between the output of the plant and model is kept small, the extent of the chattering due to the nonideal sliding motions is mitigated. While in the sliding mode, the system remains insensitive to parameter variations and disturbances. It is this insensitive property of the sliding mode control that enables the elimination of interactions along subsystems. A control input in another subsystem is fed forward to the present subsystem in order to decouple roughly. A block diagram of an improved sliding mode control scheme is shown in Fig. 3.10.

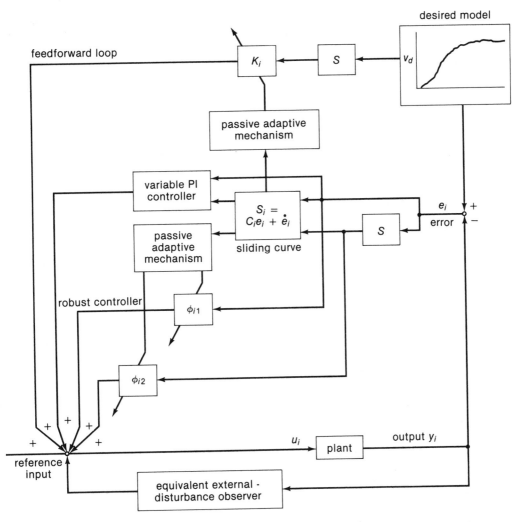

Figure 3.10   Block diagram of improved sliding mode control

Learning and Intelligent (Expert) Control

In general, it is very difficult for a motor drive system to follow the desired motion, due to the existence of mutual dynamic interference among subsystems, external disturbances, and system parameter variations. It is usually necessary to obtain knowledge of the control object in order to design controllers. To overcome these difficulties, an iterative learning control, which is one of the most important abilities of human beings, is introduced. In this iterative learning control scheme, the input of the present trial is modified by the error of the previous trial, which consists of the desired output, minus the previous response. It is reported that by using this method, the positioning error could be reduced to $\frac{1}{100}$ of the one for a conventional control method. Sliding surfaces are also used for observer design. A particular observer structure including switching terms is shown to have promising properties in the presence of modeling errors and sensor noise. Recently, a fuzzy expert system has been considered as a more promising approach than an expert system since the number of rules can be reduced in the fuzzy expert system.

Among control algorithms, practical robust control algorithms, with accurate and fast control algorithms, are important in motor (motion) control. Chapters 5 and 6 cover accurate and fast control policies and sensings.

Overview of Digital Signal Processing

In the last decade, Digital Signal Processing (DSP) has made tremendous progress in both the theoretical and practical aspects of the field. While more DSP algorithms are being discovered, better tools are also being developed to implement these algorithms. One of the most important breakthroughs in electronics technology is the high-speed digital signal processors. These single-chip processors are now commercially available in Very Large-Scale Integrated (VLSI) circuits from semiconductor vendors. Digital signal processors are essentially high-speed microprocessors/microcomputers, designed specifically to perform computation-intensive digital signal processing algorithms. By taking advantage of the advanced architecture, parallel processing, and dedicated DSP instruction sets, it can execute millions of DSP operations per second. This capability allows complicated DSP algorithms to be implemented in a tiny silicon chip, which previously required the use of a minicomputer and an array processor.

With this VLSI advancement, innovative engineers in industry are discovering more and more applications where digital signal processors can provide a better solution than their analog counterparts, for reasons of reliability, reproducibility, compactness, and efficiency. These digital signal processors are also highly programmable, which makes them very attractive for: (1) system upgrades, in the case of advancements in DSP algorithms, and (2) multitasking, where different tasks can be performed with the same device by simply changing its program. Because of these and many other advantages, digital signal processors are becoming more prevalent in areas of general-purpose digital control.

One example is the TMS 320C25 Digital Signal Processor, which is the second generation (member) of the TMS 320 family of VLSI digital signal processors and peripherals. The TMS 320 family supports digital signal processing (DSP) applications, such as spectrum analysis, digital filtering, high-speed control, and other computation-intensive applications.

With a 100-ns instruction cycle time and an innovative memory configuration, the TMS 320C25 performs operations necessary for many real-time digital signal processing algorithms. Since most instructions require only one cycle, the TMS 320C25 is capable of executing five million instructions per second. The

TMS 320C25 also contains special repeat instructions for streamlining program space and execution time. On-chip data RAM of 544 16-bit words, direct addressing of up to 64K words each of external data and program memory, and multiprocessor interfacing features for shared global memory, minimize the overhead involved in data transfers, to take full advantage of the capabilities of the processor.

Development tools and applications support are key advantages to using the TMS 320C25. Full-speed emulators, software simulators and assemblers, and extensive documentation, including application reports, provide for rapid design and development cycles. Regional technology centers, system application engineers, and third-party support are available for DSP education and design.

Digital Notch Filter Design

In order to achieve fast motion control for mechanical systems, mechanical parts are designed to be much higher in weight and smaller in size as is technically possible. Therefore, mechanical oscillation problems arise. A digital notch filter for eliminating the oscillations in mechanical parts is presented as follows:

A digital filter is a compensator or, actually, a digital controller. Since it is implemented with software (programs), the digital filter of desired performance

TABLE 3.3 ANALOG AND DIGITAL NOTCH FILTERS

Compensator/ Filter Element	Analog Transfer Function	Digital Transfer Function $(f_g = 4020 \text{ Hz})$
1800-Hz Notch Filter	$G8 = \dfrac{S^2 + (2\pi\,1800)^2}{S^2 + \dfrac{2\pi\,1800}{5}S + (2\pi\,1800)^2}$	$D8 = \dfrac{0.96877 + 1.83411Z^{-1} + 0.96877Z^{-2}}{1.0 + 1.83411X^{-1} + 0.93754Z^{-2}}$
900-Hz Notch Filter	$G9 = \dfrac{S^2 + (2\pi\,900)^2}{S^2 + \dfrac{2\pi\,900}{2.5}S + (2\pi\,900)^2}$	$D9 = \dfrac{0.8352 - 0.27291Z^{-1} + 0.8352Z^{-2}}{1.0 - 0.27291Z^{-1} + 0.67041Z^{-2}}$
560-Hz Notch Filter	$G10 = \dfrac{S^2 + (2\pi\,560)^2}{S^2 + \dfrac{2\pi\,560}{5}S + (2\pi\,560)^2}$	$D10 = \dfrac{0.9287 - 1.19021Z^{-1} + 0.9287Z^{-2}}{1.0 - 1.19021Z^{-1} + 0.8574Z^{-2}}$
140-Hz Notch Filter	$G11 = \dfrac{S^2 + (2\pi\,140)^2}{S^2 + \dfrac{2\pi\,140}{5}S + (2\pi\,140)^2}$	$D11 = \dfrac{0.97875 - 1.91083Z^{-1} + 0.97875Z^{-2}}{1.0 - 1.91083Z^{-1} + 0.95751Z^{-2}}$
120-Hz Notch Filter	$G12 = \dfrac{S^2 + (2\pi\,120)^2}{S^2 + \dfrac{2\pi\,120}{5}S + (2\pi\,120)^2}$	$D12 = \dfrac{0.9817 - 1.92896Z^{-1} + 0.9817Z^{-2}}{1.0 - 1.92896Z^{-1} + 0.96339Z^{-2}}$
100-Hz Notch Filter	$G13 = \dfrac{S^2 + (2\pi\,100)^2}{S^2 + \dfrac{2\pi\,100}{5}S + (2\pi\,100)^2}$	$D13 = \dfrac{0.98467 - 1.94534Z^{-1} + 0.98467Z^{-2}}{1.0 - 1.94534Z^{-1} + 0.96935Z^{-2}}$

(frequency characteristic) can be realized. Analog filter implementation is quite difficult since resistors and capacitors of exact values must be chosen.

A TMS 320 is suitable to digital notch filter implementation in both hardware and software.

The technique for the conversions of the analog notch filters to their digital counterparts is the bilinear transformation with frequency prewarping.

The transfer functions, shown in Table 3.3, list both the analog prototypes and their digital equivalents; this will be described in Chapter 7.

3.2 MODERN CONTROL THEORY TERMINOLOGIES

In this section, modern control including signal processing theory terminologies used in this book are illustrated in a heuristic way.

Z-Transformation

The transfer function $G(s)$ is transferred into the pulse transfer function $G(z)$ as follows:

1. Obtain

$$£^{-1}[G(s)] = g(t)$$

   by applying the Inverse Laplace Transformation Table.

2. Calculate

$$g^*(t) = \sum_{K=0}^{\infty} g(kT)1(t - kT)$$

   where T is a sampling time.

3. Obtain

$$£[g^*(t)] = G^*(s) = \sum_{K=0}^{\infty} g(kT)e^{-kTS}$$

   and substitute Z into e^{TS}.

4.
$$G(z) = \sum_{K=0}^{\infty} g(kT)z^{-K}$$

   where z^{-N} is a shift operator such that $z^{-N}f(n) = F(n - N)$. The pulse transfer function $G(z)$ may be rewritten as

$$G(z) = \frac{q(z^{-1})}{p(z^{-1})}$$

   where p, q are polynomials of z^{-1}.

The solutions of $p(z^{-1}) = 0$ with respect to z are called poles or eigenvalues. If, and only if, the magnitude of the pole is less than unity in the z domain, then $G(z)$ is stable. The solutions of $q(z) = 0$ with respect to z are named zeros. $G(z)$ is defined as an input-output representation:

$$G(z) = \frac{y(z)}{u(z)}$$

or an outer-representation. $G(z) = q(z^{-1})$ is named a moving average (MA) model, or zero model.

$$G(z) = \frac{1}{q(z^{-1})}$$

is called an auto-regressive (AR) model or a pole model.

Therefore:

$$G(z) = \frac{q(z^{-1})}{p(z^{-1})}$$

is called an auto-regressive moving average (ARMA) model.

State Equation

This representation is another form of $G(z)$, and is called "inner-representation." It is obtained by introducing the state $x(z^{-1})$.

1.
$$G(z) = \frac{y(z)}{u(z)} = \frac{y(z)}{x(z^{-1})} \frac{x(z^{-1})}{u(z)}$$

where u, y are the input and the output.
Define

$$\frac{y(z)}{x(z^{-1})} = q(z^{-1}) \quad \text{and} \quad \frac{x(z^{-1})}{u(z)} = \frac{1}{p(z^{-1})}$$

or

$$y(n) = d_1 x(n - 1) + d_2 x(n - 2) + \cdots + d_m(n - m)$$

and

$$u(n) = n_0 x(n) + n_1 x(n - 1) + \cdots + d_N x(n - N)$$

where

$$q(z^{-1}) = d_1 z^{-1} + d_2 z^{-1} + \cdots + d_m z^{-m}$$

and

$$p(z^{-1}) = n_0 + n_1 z^{-1} + \cdots + n_N z^{-N}$$

Define

$$x(n - k) = x_{N-k+1}(n) \qquad (k = 1 \text{ to } N)$$

Then the following discrete state equation is obtained:

$$\begin{bmatrix} x(n+1) \\ x_2^1(n+2) \\ \vdots \\ x_N(n+1) \end{bmatrix} = \begin{bmatrix} & 010\ldots & \\ & 001\ldots & \\ & \ldots 0 & \\ & 0\ldots 01 & \\ -\dfrac{\eta_N}{\eta_0} & -\dfrac{\eta_{N-1}}{\eta_0} \ldots & -\dfrac{n_1}{\eta_0} \end{bmatrix} \begin{bmatrix} x_1(n) \\ x_2(n) \\ \vdots \\ x_N(n) \end{bmatrix} + \begin{bmatrix} 0 \\ 0 \\ \vdots \\ \dfrac{1}{n_0} \end{bmatrix} u(n)$$

$$y(n) = [000 \ldots d_1 \ldots d\ d_m] \begin{bmatrix} x_1(n) \\ x_2(n) \\ \vdots \\ x_N(n) \end{bmatrix} + eU(n)$$

or

$$x(n + 1) = Ax(n) + Bu(n), \text{ where } g(z) = \frac{q(z^{-1})}{p(z^{-1})} + e \text{ is assumed.}$$

$$y(n) = Cx(n) + Du(n)$$

See example in Chapter 7.

$$G(z) = \frac{N_0 + N_1 z^{-1} + n_2 z^{-2}}{D_0 + D_1 z^{-1} + D_2 z^{-2}} = \frac{y(z)}{u(z)} \frac{x(z^{-1})}{x(z^{-1})}$$

$$\frac{x(z^{-1})}{u(z)} = \frac{1}{D_0 + D_1 z^{-1} + D_2 z^{-2}}$$

or

$$u(n) = D_0 x(n) + D_1 x(n - 1) + D_2 x(n - 2)$$

and

$$\frac{y(z)}{x(z^{-1})} = N_0 + N_1 z^{-1} + N_2 z^{-2}$$

or

$$y(n) = N_0 x(n) + N_1 x(n - 1) + N_2 x(n - 2)$$

$$G(z) = \frac{N_0 + N_1 z^{-1} + N_2 z^{-2}}{D_0 + D_1 z^{-1} + D_2 z^{-2}}$$

$$= \frac{N_0}{D_0} + \frac{\left(N_1 - \dfrac{D_1 N_0}{D_0}\right) z^{-1} + \left(N_2 - \dfrac{D_2 N_0}{D_0}\right) z^{-2}}{D_0 + D_1 z^{-1} + D_2 z^{-2}}$$

$$\begin{bmatrix} x_1(n+1) \\ x_2(n+1) \end{bmatrix} = \begin{bmatrix} 0 & 1 \\ -\dfrac{D_2}{D_0} & -\dfrac{D_1}{D_0} \end{bmatrix} \begin{bmatrix} x_1(n) \\ x_2(n) \end{bmatrix} + \begin{bmatrix} 0 \\ \dfrac{1}{D_0} \end{bmatrix} u(n)$$

$$y(n) = -\left[\left(N_2 - \frac{D_2}{D_0} N_0\right)\left(N_1 - \frac{D_1}{D_0} N_0\right)\right] \begin{bmatrix} x_1(n) \\ x_2(n) \end{bmatrix} + \frac{N_0}{D_0} u(n)$$

The block diagram for this representation can be easily drawn as shown in Fig. 3.11. This is a suitable expression for numerical analysis programs.

2. $G(z)$ is expanded into partial fractions, such that

$$G(z) = k_0 + \sum_{i=1}^{N} \frac{N_{1i} z^{-1}}{D_{0i} + D_{1i} z^{-1}}$$

$$\begin{bmatrix} x_1(n+1) \\ x_2(n+2) \\ \vdots \\ x_N(n+1) \end{bmatrix} = \begin{bmatrix} \dfrac{D_{11}}{D_{01}} & & & 0 \\ & \dfrac{D_{12}}{D_{02}} & & \\ & & \dfrac{D_{1N}}{D_{0N}} & \\ 0 & & & \end{bmatrix} \begin{bmatrix} x_1(n) \\ x_2(n) \\ \vdots \\ x_N(n) \end{bmatrix} + \begin{bmatrix} \dfrac{N_1}{D_{01}} \\ \dfrac{N_2}{D_{02}} \\ \vdots \\ \dfrac{N_N}{D_{0N}} \end{bmatrix} \mu(n)$$

$$y(n) = [1 \ 1 \ \ldots \ 1] \begin{bmatrix} x_1(n) \\ x_2(n) \\ \vdots \\ x_N(n) \end{bmatrix} + K_0 u(n)$$

The block diagram for this representation is shown in Fig. 3.12. The transformation from $G(z)$ to the state equation is called realization. It is not unique; however, the reverse transformation is unique and is called the minimum realization under the condition that the system is controllable and observable (pole-zero cancellation is not satisfied).

Controllability and Observability

A plant in controllable with the plant input can be used to transfer the plant from any initial state to any arbitrary state in a finite time.

A plant is observable if the initial state $x(n_0)$ can be determined uniquely when the given output $y(n)$ for $n_0 < n < n_1$ for any $n_2 > n_0$.

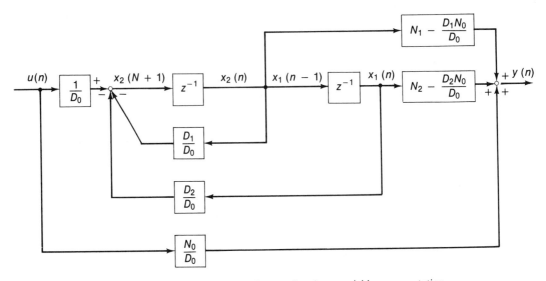

Figure 3.11   Block diagram for phase variable representation

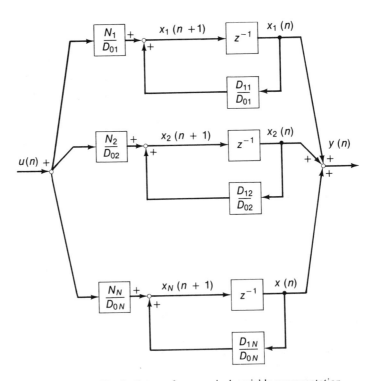

Figure 3.12   Block diagram for canonical variable representation

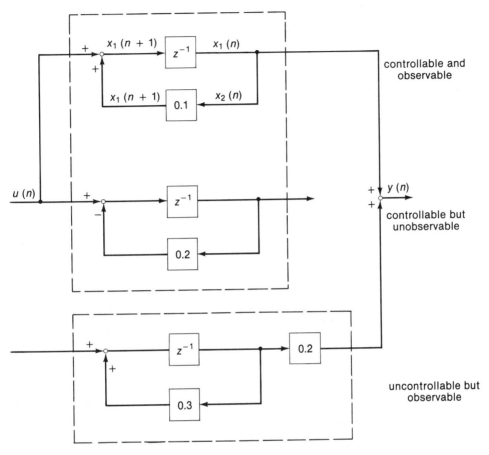

Figure 3.13   (Un)controllable and (Un)observable systems

Figure 3.13 shows (un)controllable and (un)observable systems. Controllability and observability concepts can be obtained at a glance.

Model Reference Techniques

Model reference techniques are used for identification, estimation, and control, as shown in Fig. 3.14. This concept is very close to following nonlinear system linearization.

Let a nonlinear system be

$$\dot{x} = f(x, t, u) \quad \text{(nonlinear vector differential equation → state equation)}$$

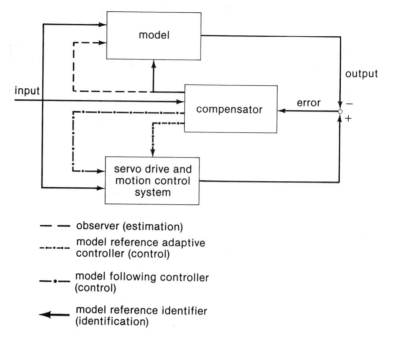

Figure 3.14 Model reference techniques

1. Obtain the nominal trajectory (xo, uo) solving optional control problems by minimizing PI.

$$PI = \min \int_0^T L(x, t, u) \, dt$$

under the condition that

$$g(x_0, t_0, u_0) = 0$$
$$\dot{x}_0 = f(x_0, u_0, t_0)$$

or by simulation obtain the nominal trajectory.

2. Then, in the vicinity of the nominal trajectory, linearize the nonlinear system by calculating Jacobian matrices as follows:

$$\Delta x = A(t)\Delta x + B(t)\Delta u(t)$$
$$\Delta y(t) = c(t)\Delta x(t)$$

where

$$A(t) = \left.\frac{\partial f}{\partial x}\right|_{\substack{u = u_0 \\ x = x_0}} \qquad B(t) = \left.\frac{\partial f}{\partial u}\right|_{\substack{u = u_0 \\ x = x_0}}$$

$$C(t) = \left.\frac{\partial y}{\partial x}\right|_{\substack{u = u_0 \\ x = x_0}}$$

The linearized system behaves like a model following control system, since its error is so small that the nonlinear system is controlled in the vicinity of the model. The input to the actuator is small enough that saturation may not be considered.

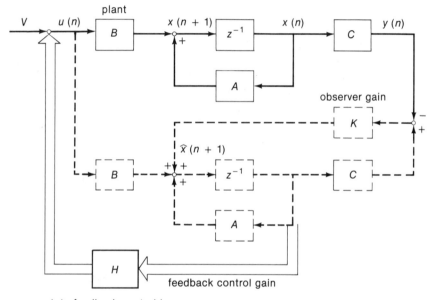

state feedback control loop

observer: $\hat{x}(n + 1) = A\hat{x}(n) + B u(n)$
$+ KC[\hat{x}(n) - x(n)]$
error; $e(n) = \hat{x}(n) - x(n)$
$e(n + 1) = (A + KC)e(n)$

all eigenvalues of $(A + KC)$ should have their magnitudes less than unity for convergence (stability).

Figure 3.15 Identity observer and state feedback control

Observer (Estimation)

An observer reconstructs the physical variable (state) unmeasurable with the sensor. The observer is designed by using a model (plant) reference technique as shown in Fig. 3.15. The observer is actually a digital filter.

State Feedback Control

The reconstructed state can be feedback to the input as shown in Fig. 3.15 to optimize the controlled system (for example, pole placement, model matching, etc.). The observer may be considered as a plant simulator except that a loop including k is provided such that

$$\lim_{n \to \infty} e(n) = 0$$

Identification (Modeling)

If robust control and estimation (described in chapters 5 and 6) are used, then this identification is not necessarily considered. However, MRAC needs this mechanism.

A model is derived on the basis of the measured input, output, and other physical variables (nonphysical model), or by applying physical law such as Newton's and Kirchhof's laws (physical model). Modern modeling takes the former one (identification) as follows:

Identification by least square method

Let

$$z(k) \triangleq \text{measured value}$$

$$y(k) \triangleq \text{true output}$$

$$v(k) \triangleq \text{noise}$$

$$z(k) = y(k) + v(k)$$

$$y(k) = \sum_{m=0}^{M} h(m)u(k - m) \qquad \text{(MV model)}$$

$h(m)$ is determined from $z(0), z(1), \ldots , z(N)$

$$\begin{bmatrix} z(0) \\ z(1) \\ \vdots \\ z(N) \end{bmatrix} = \begin{bmatrix} u(0) & u(-1) & \cdots & u(-M) \\ u(1) & \cdots & \cdots & u(1-M) \\ \vdots & & & \vdots \\ u(N) & \cdots & \cdots & u(N-M) \end{bmatrix} \begin{bmatrix} r(0) \\ r(1) \\ \vdots \\ r(M) \end{bmatrix} + \begin{bmatrix} v(0) \\ v(1) \\ \vdots \\ v(N) \end{bmatrix}$$

or

$$z = Uh + v$$
$$z = Uh$$

by using a right-hand pseudoinverse matrix

$$h = (u^T v)^{-1} v^T z$$

This can be calculated by using DSP; when modeling, the following should be taken into account:

1. Make the input and the output clear.
2. The model should be simple but express only the plant essence.
3. An identification technique for the model should be well established.

Brushless Servo Motor Model (Physical Model)

By applying the *dq* transformation

$$\begin{bmatrix} vd \\ vq \end{bmatrix} = \begin{bmatrix} Ra + PLa & -W_m La \\ W_m La & Ra + PLa \end{bmatrix} \begin{bmatrix} id \\ iq \end{bmatrix} + \begin{bmatrix} W_m \Phi \\ 0 \end{bmatrix}$$

$$Te = \tfrac{3}{4}. (\# \text{ of poles}). \ \Phi_{id} = J\dot{W}_m + BW_m + Te$$

$$Te \triangleq \text{motor torque}$$
$$La \triangleq \text{armature inductance}$$
$$Ra \triangleq \text{armature resistance}$$
$$J \triangleq \text{motor inertia}$$
$$\Phi \triangleq \text{flux (constant)}$$
$$W_m \triangleq \text{motor speed}$$
$$B \triangleq \text{viscous friction coefficient}$$
$$p \triangleq d/dt$$
$$vd, id \triangleq d\text{-axis voltage and current, respectively}$$
$$vq, iq \triangleq q\text{-axis voltage and current, respectively}$$

Two-Degrees-of-Freedom Design

Assume that plant parameters are known by identification or by equivalent disturbance cancellation; performed by zeroing and a disturbance observer, the following two-degrees-of-freedom design method is applicable:

Let's consider the control system as shown in Fig. 3.16.

1. Obtain the desired transfer function by normalizing:

$$\frac{Y_m(s)}{D_m(s)} = \frac{1}{Y_m} \frac{1}{\alpha_0 + \alpha_1 \sigma S + \alpha_2 \sigma^2 S^2 + \cdots}$$

$$\{\alpha_i\} = [1, 1, 0.5, 0.15, \ldots]$$

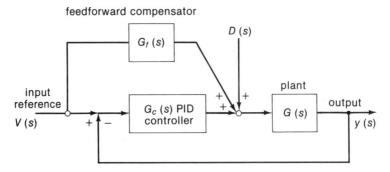

Figure 3.16 Controller and system

2. Calculate the actual transfer function:

$$\frac{Y(s)}{D(s)} = \frac{G(s)}{1 + G_c(s)G(s)} = \frac{1}{T} \frac{1}{\alpha_0 + \alpha_1 \bar{\sigma}S + \alpha_2 \bar{\sigma}^2 S^2 + \cdots}$$

3. Put:

$$\frac{Y_m}{D_m} = \frac{Y(s)}{D(s)}$$

Determine $G(s)$ parameters.

4. Obtain the desired transfer function by normalizing:

$$\frac{Y_m(s)}{U_m(s)} = \frac{1}{T'_m} \frac{1}{\alpha_0 + \alpha_1 \acute{\sigma}S + \alpha_2 \acute{\sigma}^2 + \cdots}$$

5. Calculate the actual transfer function:

$$\frac{Y(s)}{U(s)} = \frac{1}{T'} \frac{1}{\alpha_0 + \alpha \acute{\sigma}S + \alpha_2 \acute{\sigma}^2 S^2 + \cdots}$$

6. Put:

$$\frac{Y_m(s)}{U_m(s)} = \frac{Y(s)}{U(s)}$$

$G_c(S)$ for a controller to a disturbance and $G_f(s)$ for a controller to a reference input can be designed independently. This design concept contains a partial model matching design concept.

Stability Theory

Stability is the necessary condition for identification, estimation, and control problems, since they are reduced to the stability problems of error dynamic equations.

To prove theories, the following Liapunov Stability method is useful:

Let a nonlinear system be

$$x(n + 1) = f[x(n)]$$

and a Liapunov function be $V(n)$, where $V(n)$ is a positive definite function with respect to $x(n)$. For stability:

$$V(n + 1) - V(n) < 0$$

BIBLIOGRAPHY

[3.1] Y. Dote, "Application of Modern Control Techniques to Motor Control," Proc. of the IEEE, April 1988.

Chapter 4

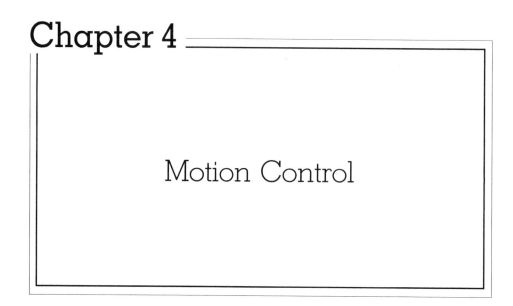

Motion Control

4.0 INTRODUCTION

Electromechanical motion control systems have changed dramatically in the past decade since revolutionary progress in power electronics and microelectronics has been made.

By using a servo motor drive system developed with interdisciplinary engineerings (mechanical and electronic engineerings), it has become possible to control at will the position, velocity, acceleration, and force of a mechanical system in the space. This technology has contributed to providing high productivity on the production line and to producing high-quality products which are the basis of modern industry development. Advanced motion control requires high speed, large allowable current, reliable semiconductor switches, fast processors (DSPs), advanced control, and sensor signal processing algorithms described in chapters 5 and 6. This chapter describes important problems in motion control.

4.1 POSITIONING ACCURACY AND SPEED CONTROL RATE

There are two kinds of controls in positioning. One is called PTP (point to point control) and the other is CPC (continuous path control). In both cases, considerations in accuracy, response time, and cost are required. The choice of sensors is the most important factor in obtaining positioning accuracy; however, even when an accurate sensor is used, an overall system is necessary to yield high resolution in positioning control.

The speed control rate and the control gain in the servo control loop play an important role in high resolution positioning. Assume that a sensor is ideal, no time delay in the speed control loop exists, and there is a nonlinearity, shown in Fig. 4.1, between the command input and the output. The servo motor will not move for a small input. In other words, the servo motor will not rotate, and will yield uncertain behavior (oscillation or hysteresis rotation), if the command input stays within a. The input range within a is called a dead band. On the other hand, the maximum allowable speed is limited to V_{0max} in the high-speed operation. The command output is saturated. In general, V_{0max}/a is defined as the speed control ratio. As shown in Fig. 4.1, the input and the output are normalized such that $v_0/v_i = 1$. The loop gain in position control loop is represented by the preamplifier gain, K. V_i should be at least greater than a for the motor to start rotating from its stalled position. Therefore, the required speed control ratio N is given by $N > = V_{0max}/a = V_{0max}/ek$. Let the resolution be 10, K be 30, and V_{0max} be 120 mm/sec. Then N should be greater than or equal to 400 (jerking motion). Allowing some margin, N must be between 800 and 1000. Therefore, the increase in K results in not only fast dynamic response, but also accurate positioning.

Next, let us consider the tracking error in continuous path (CP) control. The trajectory in the x-y plane produces the tracking error shown in Fig. 4.2 due to the delayed output in the control system. As the following equation is satisfied:

$$y = Vy/Vx * x - Vy(ey - ex), \Delta y = .Vy(ey - ex) \qquad (4.1)$$

The more complicated CP becomes, the more error CP control gives.

Feedforward control is effective in canceling this error. This is discussed in Chapter 5.

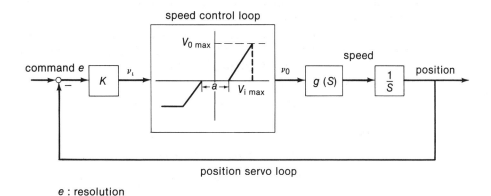

Figure 4.1   Required speed control ratio

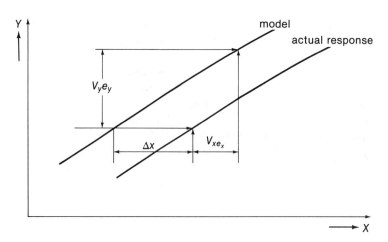

Figure 4.2 Tracking error in *x-y* plane

4.2 FORCE AND ACCELERATION CONTROL

Force Control

There are current and speed control loops in an available servo drive system. In order to achieve force control by using this servo drive, it is necessary to identify the compliance of the controlled object, and then to change the feedback gain according to the identified compliance. This control block diagram is shown in Fig. 4.3. In short, an adaptive gain control is required. It is very difficult to grip stably every object from a soft tennis ball to a hard steel ball with a commercially available servo drive system.

In this section, a novel intelligent control algorithm is introduced. The control consists of variable control gains whose determinations are based on the fuzzy set theory with some knowledge engineering (expert system) and the improved sliding mode control method.

It is known from experience that the control gain K_f is the reciprocal to the compliance K_e. This K_e is determined by using the experiment expressed by a set of fuzzy membership functions F_1 and F_2.

First, K_e is obtained as follows:

$$K_e(k) = [F(k) - F(k - 1)]/V(k - j)$$

Since

$$K_e = dF/dx = (dF/dt)/(dx/dt) = F/V$$

where $F \triangleq$ measured force, $V \triangleq$ speed command, and $j \triangleq$ servo system delay time.

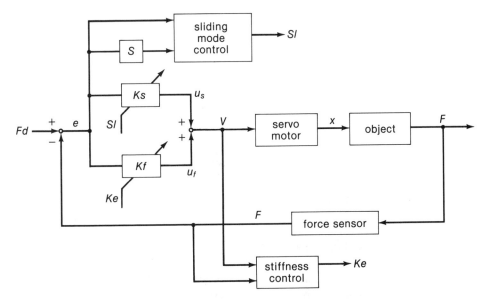

Figure 4.3 Control block diagram

Then, membership functions F_1 and F_2 are determined by experience. F_1 is for a soft tennis ball. F_2 is for a rigid steel ball. They are shown in Fig. 4.4.

$$W_{\max} = F_1(K_e)$$

$$W_{\min} = F_2(K_e)$$

$K_{e\max}$ (0.2) for a rigid steel ball and $K_{e\min}$ (0.02) for a soft tennis ball are obtained by experiment.

Finally, decision making (obtaining K_e) is achieved by calculating the fol-

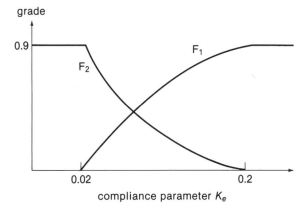

Figure 4.4 Membership functions

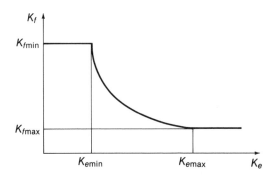

Figure 4.5 Variable gain versus estimated compliance

lowing equation:

$$K_e = (K_{emax}W_{max} + K_{emin}W_{min})/(W_{max} + W_{min})$$

The obtained gain $(1/K_e)$ is shown in Fig. 4.5.

Next, an improved sliding mode controller is applied in parallel with the developed fuzzy set controller. The goal is to find the variable feedback control gain K_S to nullify the error e as t goes to infinity.

$$K_e = sl_i/(|sl_i| + d_i)$$

where $d_i = 1.$, $sl_i = CE + e$, and C is a constant. Sl_i is called a sliding curve.

The developed control algorithm is implemented with a fast digital signal processor. This processor is also used for implementing a moving average and variable structure digital filters, in order to cancel noise from the sensors.

Another force control method is performed well by using zeroing techniques described in Chapter 5.

Acceleration Control*

By applying zeroing with an equivalent disturbance observer, ω_m, a servo motor control system has nominal parameters as shown in Fig. 4.6. Therefore, an exact acceleration reference can be applied where:

$$W_m \triangleq \text{motor angular velocity}$$

$$\dot{W}^{ref} \triangleq \text{motor angular acceleration reference}$$

$$J_n \triangleq \text{nominal inertia}$$

$$K_{tn} \triangleq \text{nominal torque constant}$$

$$\theta \triangleq \text{motor angular displacement}$$

* See reference 4.2.

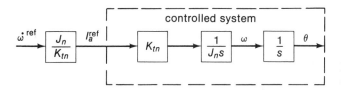

Figure 4.6 Servo motor system after equivalent disturbance cancellation

4.3 CONTROL CONSIDERATIONS IN MECHANICAL SYSTEMS WITH NONLINEARITIES

Even if a high-performance servo motor and its driver are used, high control performance cannot be achieved when there are nonlinearities, such as friction hysteresis (gear backlash) and compliance in the mechanical parts of the overall system. In order to cancel the effects of these nonlinearities, robust mechanical parts are designed. Then, servo control performance becomes worse, due to the resultant heavier mechanical loads. Appropriate sensors and controllers can solve this problem as follows:

1.   Friction control. If the transferring torque is saturated and the force transmitting part has slipped, then overshoot is observed in the system dynamic response. To cope with this, it is necessary to adjust the driver speed to the load speed by sensing the slip.

2.   Compliance control. By feeding back the displacement of the compliance object to the control input, the oscillatory system can be damped. This also makes the resonant (natural) frequency of the mechanical part higher.

3.   Backlash control. This control is achieved by the following procedure:

a.   By using the trajectory in the phase plane. Whether the load should be accelerated or deccelerated should be determined.

b.   Whether the load touches the driving parts in the accelerating or deccelerating direction should be determined by sensors mounted on the driving motor and the driven load.

c.   In the case when the load touches in either direction, the load position should be controlled.

d.   If the load does not touch in any direction, the gap between the load and the driving part should be adjusted so that the load may be touched at zero mutual speed.

4.4 MECHANICAL IMPEDANCE CONTROL

First, force control as described in section 4.2 is considered. As discussed in section 4.2, a servo motor drive system has nominal parameters. Accurate force control can be achieved as shown in Fig. 4.7, where Dn is a nominal damping coefficient.

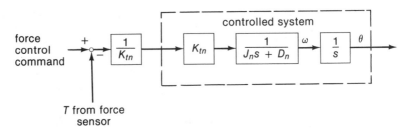

force
control
command

T from force
sensor

Figure 4.7   Force control

By integrating $\dot{W}^{\text{ref}}$ in Fig. 4.6, accurate angular velocity command and accurate angular displacement command can be applied to the system, as shown in Fig. 4.8.

The following equation is obtained by applying Newton's motion rule:

$$T(s) = (s^2 J + sD) + K)\theta(s) \qquad (4.2)$$

$(s^2 J + sD + K)$ is called a mechanical impedance, or

$$s^2\theta_c(s) = [T(s) - (D_c S + k_c)\theta_c(s)]/J_c \qquad (4.3)$$

where $(s^2 J_c + sD_c + K_c)$ is a target impedance.

Then $s^2\theta_c(s)$ is the acceleration command.

By integrating $s^2\theta_c(s)$, the angular velocity command $s\theta_c(s)$ and the angular displacement command $\theta_c(s)$ can be obtained. They are applied to the inputs of the system shown in Fig. 4.8. This is called mechanical impedance control.

Figure 4.9 shows mechanical impedance control for a multi-degrees-of-free-

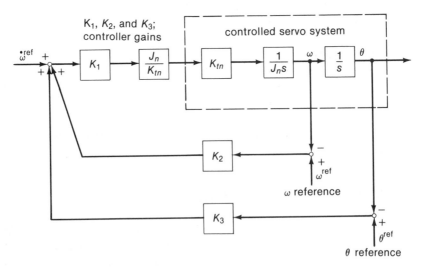

Figure 4.8   Robust and accurate system in which $\dot{W}^{\text{ref}}$, ω^{ref}, and θ^{ref} are applied

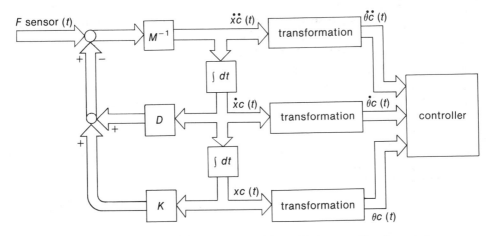

Figure 4.9   Mechanical impedance control for multi-degrees-of-freedom robot

dom robot, where M is an inertia or mass matrix, D is a viscous friction coefficient matrix, and k is a compliance matrix.

4.5 MECHANICAL VIBRATION SUPPRESSION CONTROL

Modern computer-integrated manufacturing systems require quick motion in mechanical systems powered by motors in order to obtain high productivity. Therefore, mechanical systems are designed so that the weight is minimal. Consequently, mechanical parts, especially gear trains and torque transmission mechanisms, have compliances. Rigid mechanical parts give large inertias. Thus, vibration problems due to these compliances and inertias arise in high-speed operation of mechanical systems. Usually, mechanical systems yield nonlinear problems. The following vibration characteristics should be considered:

1. Natural vibration frequency is variable due to load variations and mechanism attitude changes.
2. Multi-vibration frequencies are observed due to interactions among mechanisms.

Generally, the following methods are proposed for suppressing vibrations:

1.   Mechanical method. Mechanical dampers are used to absorb vibrations; however, this method is not robust. It is sensitive to system parameter variations and the damper is costly to attach to the mechanism.

2.   Notch filter. A digital notch filter can be implemented with DSP's to cancel vibration at the natural frequency, but it is not robust when there are variations in the natural frequency.

3. Smooth speed command. Slow acceleration and decceleration is given by the smooth speed command in order to avoid vibrations. However, the method of generating speed commands is complicated.

4. Force, compliance feedback control. Presently, this method is most used. Force or compliance is measured or estimated, then it is fed back to the input, but it is expensive to implement. Therefore, the following state observer method is recommended (see reference [4.1]).

Figure 4.10 shows a mechanical system which exhibits vibration. This system is described by the following equation:

$$J_M\ddot{\theta}_M + D_M\dot{\theta}_M + \frac{K_c n}{R_G}\theta_s + \frac{D_{L2} n}{R_G}\dot{\theta}_s = K_T n I_A$$

$$J_L\ddot{\theta}_L + D_{L1}\dot{\theta}_L - K_c\theta_s - D_{L2}\dot{\theta}_s = 0$$

$$\theta_s = \frac{\theta_M}{R_G} - \theta_L \tag{4.4}$$

where $\theta_M \triangleq$ motor angular position

$\theta_L \triangleq$ load angular position

$\theta_S \triangleq$ spring angular position

The current loop equation is

$$L\dot{I}_A + (R + K_A K_{AF})I_A = K_{AU} - K_E\dot{\theta}_M \tag{4.5}$$

Motor inductance L is small enough to neglect the electrical time constant such that

$$L/(C_r + K_a K_{af}) = 0 \tag{4.6}$$

$$I_a = (K_a u - K_{em})/R_e \tag{4.7}$$

where

$$R_e = R + K_a K_{af}$$

Figure 4.11 shows a control block diagram.

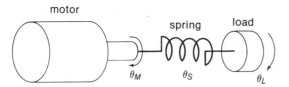

motor

spring   load

θ_M θ_S θ_L

Figure 4.10 Mechanical system

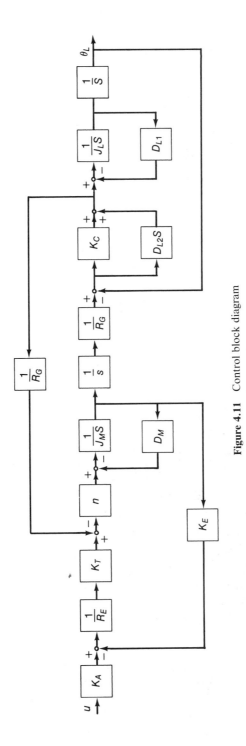

Figure 4.11 Control block diagram

The system state vector is defined as follows:

$$X \triangleq [\dot{\theta}_M \theta_M \dot{\theta}_S \theta_S]^T \qquad (4.8)$$

θm and θm (the output vector y) are measurable.

$$Y \triangleq [\dot{\theta}_M \theta_M]^T \qquad (4.9)$$

The following state equations are derived from Eqs. (4.4) through (4.6):

$$\dot{X} = AX + bu \qquad (4.10)$$

$$Y = CX \qquad (4.11)$$

where

$$A = \begin{bmatrix} a_{11} & 0 & a_{13} & a_{14} \\ 1 & 0 & 0 & 0 \\ a_{31} & 0 & a_{33} & a_{34} \\ 0 & 0 & 1 & 0 \end{bmatrix}, \quad B = \begin{bmatrix} b_1 \\ 0 \\ b_3 \\ 0 \end{bmatrix}, \quad C = \begin{bmatrix} 1000 \\ 0100 \end{bmatrix}$$

$$a_{13} = -\frac{nD_{L2}}{J_M R_g}, \quad a_{14} = -\frac{nK_c}{J_M R_G} \qquad a_{34} = -\left(K_c/J_L + \frac{nK_c}{J_M R_G^2} \right)$$

$$a_{31} = \frac{D_{L1}}{J_L R_G} - \frac{1}{J_M R_G}\left(D_M + \frac{nK_E K_T}{R_E} \right) \qquad b_1 = nK_A K_T/(J_M R_E)$$

$$a_{33} = -\left(\frac{D_{L1} + D_{L2}}{J_L} + \frac{nD_{L2}}{J_M R_G^2} \right) \qquad b_3 = nK_A K_T/(J_M R_E R_G)$$

State feedback with observer:

The conventional feedback control u is given in Eq. (4.12).

$$U = K_p \theta_R - [K_v K_p] Y \qquad (4.12)$$

where $K_p \triangleq$ position control loop gain
$K_v \triangleq$ velocity control loop gain
$r \triangleq$ position control command

Figure 4.12 shows complex pole locations related to gains K_v and K_p.

The adjustment of the gains K_v and K_p alone cannot stabilize the system. This is because the actual system is of a higher order, i.e., the system is not being modeled correctly.

Thus, the following state feedback control is proposed:

$$u = K_{pr} - [K_v K_p K_{vs} K_{ps}] \qquad (4.13)$$

Figure 4.13 shows the pole locations assigned by the state feedback control. The minimum order state observer is designed to reconstruct the states.

The Jacobian matrix A in Eq. (4.10) is partitioned as follows:

$$A \triangleq \begin{vmatrix} A_{22} & A_{21} \\ A_{12} & A_{11} \end{vmatrix} \quad A_{11}, A_{12}, A_{21}, A_{22}: 2 \times 2 \text{ matrix}$$

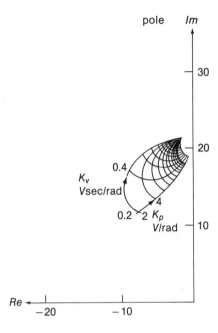

Figure 4.12 Pole placement by output feedback

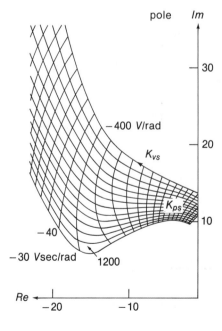

Figure 4.13 Pole locations assigned by state feedback control

Equation (4.10) is transformed linearly as follows:

$$v = TAT^{-1}v + TBu$$

$$= \begin{bmatrix} -KA_{21} + A_{11} & KA_{22} + A_{12} - (KA_{21} + A_{11})K \\ A_{21} & A_{22} - A_{21}K \end{bmatrix} v + \begin{bmatrix} K & I_{22} \\ I_{22} & 0 \end{bmatrix} Bu$$

$$(4.14)$$

The upper half portion of the above equation represents the observer. It is given in Eqs. (4.15) and (4.16).

$$\dot{\hat{z}} = F\hat{a} + Gy + WBu \qquad (4.15)$$

$$\hat{x} = T^{-1}v = Hz + jY \qquad (4.16)$$

where

$$\hat{x} \triangleq [\theta_M \theta_M \hat{\dot{\theta}}_s \hat{\theta}_s]^T \qquad\qquad W = [KI_{22}]$$

$$\hat{z} \triangleq Ky + [\hat{\dot{\theta}}_s \hat{\theta}_s]^T \qquad\qquad H = \begin{bmatrix} 0_{22} \\ I_{22} \end{bmatrix}$$

$$F = KA_{21} + A_{11}$$

$$G = A_{12} + KA_{22} - FK \qquad\qquad J = \begin{bmatrix} I_{22} \\ -K \end{bmatrix}$$

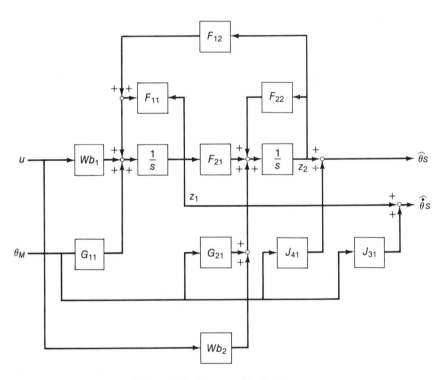

Figure 4.14 Observer block diagram

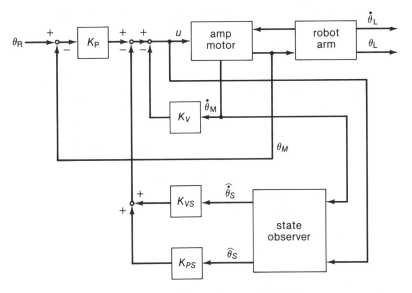

Figure 4.15   Overall control system block diagram

The matrix K is arbitrarily determined, but the first columns of the matrix, K_{11} and K_{21}, can be used for pole assignment (two poles).

The second column of the K matrix, K_{12} and K_{22}, can be made zero so as to eliminate noise contained in the measured θ_m. That also makes the matrix operation easy. If K_a and K_{af} are large, and Eq. (4.7) is satisfied, then the estimate is not very biased.

Figure 4.14 shows a block diagram for the observer. Figure 4.15 shows the block diagram for the overall control system with the constructed observer. This vibration suppression observer is implemented with a DSP. A nonlinear observer is also applicable to suppress this kind of vibration.

An equivalent disturbance observer is effective to deprive a flexible system of mechanical vibration. In section 4.4, the increased target damping coefficient D_c stabilizes the oscillatory system.

It is noted that an equivalent disturbance observer is good and effective enough to suppress mechanical vibrations (see reference [4.2]).

BIBLIOGRAPHY

[4.1] Takao Wada, et al., "Vibration Suppression Control for Robots by Software," Transactions of the IEEE Japan, vol. 107-D, No. 8, 1987.

[4.2] K. Ohnishi, et al., "Robust Force and Compliance Control of Robotic Manipulator," proc. of the IECON'88, Singapore, 24–28, Oct. 1988.

Section II
Digital Control and Signal Processing

Chapter 5

Digital Robust Control Algorithms

5.0 INTRODUCTION

This chapter covers robust digital control algorithms with which advanced motion control can be expected.

The control algorithms are novel, promising, and are explained by giving many examples; perhaps they also eliminate a gap that has existed between control theory and practice. Since they are robust, it is easy to determine controller parameters (easy controller design). A simple (fuzzy) expert system can be effectively applicable to controller design using a DSP.

They are applicable not only to a single-loop servo control system, but also to a large-scale system, since their robustness cancels the equivalent disturbances from another subsystem. In other words, they provide decoupling capability, and are also effective to control a direct drive (without gear trains) servo motor, which gives highly nonlinear properties. This is also due to its decoupling capability. Finally, they provide servo drives with efficient, accurate, and fast control performance.

5.1 EQUIVALENT DISTURBANCE OBSERVER AND ZEROING

This method is explained from three different points of view.

Digital 2-Delay Zeroing with Observer*

1. Equivalent Disturbance Observer

a. Discretization
A continuous dynamical system can be represented as follows:

$$x(t) = A_c x(t) + b_c u(t) + h_c d(t) \tag{5.1}$$

$$y(t) = C_c x(t) \tag{5.2}$$

where u: control $\in R$, y: output $\in R$, $x \in R^n$, A_c, b_c, h_c are known constant matrices with the appropriate sizes, and d is the equivalent disturbance, including parameter disturbances.

By using a sampler and holder of order-zero:
Equation (5.1) is discretized:

$$x_{i+1} = Ax_i + bu_i + hd_i \tag{5.3}$$

where

$$y_i = CX_i \tag{5.4}$$

The notation $x_{i+1} = x(i + 1)$, etc.

$z_i = y(iT + mT)$ is defined as the output between the sampling instant iT, and mT $(0 < m < 1)$, which is assumed to be constant, then

$$z_i = c\hat{A}x_i + c\hat{b}u_i + c\hat{h}d_i \tag{5.5}$$

where

$$A \exp(A_c T), \qquad b = \int_0^T \exp(A_c \iota) \, d\iota b_c$$

$$h = \int_0^T \exp(A_c \iota) \, d\iota h_c, \qquad c = c_C, \qquad \hat{A} = \exp(A_c mT)$$

$$\hat{b} = \int_0^{m\iota} \exp(A_c \iota) \, d\iota b_c \quad \text{and} \quad \hat{h} = \int_0^{m\iota} \exp(A_c \iota) \, d\iota h_c$$

b. Equivalent Disturbance Observer
From Eq. (5.5):

$$d_i = (c\hat{h})^{-1}(z_i - c\hat{A}x_i - c\hat{b}u_i) \tag{5.6}$$

In Eq. (5.6), z is observable; therefore, the disturbance estimate $\hat{d}_i$ can be obtained by using the estimate $\hat{x}_i$.

$$\hat{d}_i = (c\hat{h})^{-1}(z_i - c\hat{A}\hat{x}_i - c\hat{b}u_i) \tag{5.7}$$

Next, x_i can be formed by using a state observer.

* See reference [5.1].

Substituting Eq. (5.6) into Eq. (5.3):

$$x_{i+1} = (A - h(c\hat{h})^{-1}c\hat{A})x_i + h(c\hat{h})^{-1}z_i + (b - h(c\hat{h})^{-1}c\hat{b})u_i \qquad (5.8)$$

Since if (c, A) is an observable pair, then it is known that $(c, A - h(c\hat{h})^{-1}c\hat{A})$ is also an observable pair, an identity state observer for Eq. (5.8) is constructed as follows by using an observer gain k.

$$\hat{x}_{i+1} = (A - h(c\hat{h})^{-1}c\hat{A} - k_c)\hat{x}_i + ky_i + h(c\hat{h})^{-1}z_i + (b - h(c\hat{h})^{-1}c\hat{h})u_i$$

$$(5.9)$$

Subtracting Eq. (5.8) from Eq. (5.9), the error difference equation with regard to the error estimate $e_i = \hat{x}_i - x_i$ is derived.
Such that

$$e_{i+1} = (A - h(c\hat{h})^{-1}c\hat{A} - k_c)e_i \qquad (5.10)$$

The equivalent disturbance estimate error ζ_i defined as:

$$\zeta_i = \hat{d}_i - d_i = -(c\hat{h})^{-1}c\hat{A}e_i \qquad (5.11)$$

By choosing k in Eq. (5.10) so that Eq. (5.10) can be stabilized, e approaches zero as time goes to infinitive. Consequently, ζ_i in Eq. (5.11) converges to zero.
   c.   Convergence Improvement of Estimate Error Equation
   An overshoot will be observed in the time response of Eq. (5.10). In order to reduce this overshoot, a compensation term representing $\hat{y}_i = cx_i$ and y_i are added in Eq. (5.7). Then, Eq. (5.7) becomes Eq. (5.12).

$$\hat{d}_i = (c\hat{h})^{-1}(z_i - c\hat{A}\hat{x}_i - c\hat{b}u_i) - \propto(c\hat{h})^{-1}(\hat{y} - y_i) \qquad (5.12)$$

Substracting Eq. (5.6) from Eq. (5.12):

$$\zeta_i = \hat{d}_i - d_i$$

becomes

$$\zeta_i = (c\hat{h})^{-1}(\propto c - c\hat{A})(\hat{x}_i - x_i) = (ch)^{-1}(\propto c - c\hat{A})e_i \qquad (5.13)$$

$\propto$ is chosen, such that the norm of $\zeta_i \mid \zeta_i \mid$ is minimized.

$$\mid \zeta_i \mid \leq \mid (c\hat{h})^{-1}(\propto c - c\hat{A})\mid \mid e_i \mid \qquad (5.14)$$

where $\mid \mid$ is defined as the l_2 norm.
   Cost function J is defined as

$$J = \mid (c\hat{h})^{-1}(\propto c - cA)\mid^2 \qquad (5.15)$$

when

$$\propto = (cc^T)^{-1}(c\hat{A}c^T) \qquad (5.16)$$

J takes the minimum value J when

$$J_{min} = \mid (c\hat{h})^{-1}(cc^T)^{-1}(c\hat{A}c^Tc - cc^Tc\hat{A})\mid^2 \qquad (5.17)$$

2. Equivalent Disturbance Cancellation by Zeroing

If b_c and h_c in Eqs. (5.11) and (5.12) are linearly dependent, then by feed-forwarding $\hat{d}_i$, the equivalent disturbance can be cancelled. Otherwise, zeroing is applied to d_i. At least the condition that $ch = 0$ in general is necessary for zeroing d in Eqs. (5.3) and (5.4). However, in general, $ch \neq 0$; therefore zeroing cannot be applied.

In order to avoid the limitation that $ch \neq 0$ and the system is a nonminimum phase system, the observable output between the Fig. 5.1 sampling instances can be used as follows in Eq. (5.1):

$$x_{i+1} = Ax_i + \check{A}\hat{b}v_i + \hat{b}\omega_i + hd_i \tag{5.17a}$$

$$y_i = cx_i \tag{5.17b}$$

$$z_i = c\check{A}x_i + c\hat{b}v_i + c\hat{h}d_i \tag{5.17c}$$

where

$$\check{A} = \exp(A_c(1 - m)T), \qquad \hat{b} = \int_0^{m\iota} \exp(A_c\iota)\,d\iota b_c$$

Calculating $\hat{d}_i$ and $\hat{x}_{i+1}$ for Eq. (5.17) in the same way used as the one in the previous section:

$$\hat{d}_i = (c\hat{h})^{-1}(z_i - c\hat{A}\hat{x}_i - c\hat{b}v_i) \tag{5.18}$$

$$\hat{x}_{i+1} = (A - h(c\hat{h})^{-1}c\check{A} - k_c)\hat{x}_i + ky_i + h(c\hat{h})^{-1}z_i$$
$$+ (\check{A}\hat{b} - h(c\hat{h})^{-1}e\hat{b})v_i + \check{b}u_i \tag{5.19}$$

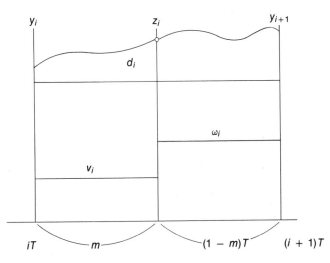

T; sampling time

Figure 5.1   2-delay sampling

The same error Eqs. (5.10) and (5.11) can be obtained for Eqs. (5.19) and (5.18).

Since $\hat{d}_i$ is obtained from Eqs. (5.18) and (5.19), for Eq. (5.17):

$$\omega_i = -(c\breve{b})^{-1}ch\hat{d}_i + u_i \tag{5.20}$$

is feedback, where u_i is the new input.

By substituting Eq. (5.20) into Eq. (5.17a) and using Eq. (5.11), d_i can be cancelled as follows:

$$x_{i+1} = Ax_i + \breve{b}(c\breve{b})^{-1}ch(c\hat{h})^{-1}c\hat{A}e_i + \breve{A}\breve{b}v_i + \breve{b}u_i + (h - \breve{b}(c\breve{b})^{-1}ch)\,d_i \tag{5.21}$$

The following augmented system can be written from Eqs. (5.10), (5.21), and (5.17b):

$$\tilde{x}_{i+1} = \tilde{A}\tilde{x}_i + \tilde{b}_1 v_i + \tilde{b}_2 u_i + \tilde{h}\,d_i \tag{5.22a}$$

$$y_i = \tilde{c}\tilde{x}_i \tag{5.22b}$$

where

$$\tilde{A} = \begin{bmatrix} A, & \breve{b}(c\breve{b})ch(c\hat{h})^{-1}c\hat{A} \\ 0, & A - h(ch)^{-1}cA - k_c \end{bmatrix}$$

$$\tilde{h} = \begin{bmatrix} h - \breve{b}(c\breve{b})^{-1}ch \\ 0 \end{bmatrix}$$

$$\tilde{x}_i = \begin{bmatrix} x_i \\ e_i \end{bmatrix}, \qquad \tilde{b}_1 = \begin{bmatrix} Ab \\ 0 \end{bmatrix}, \qquad \tilde{b}_2 = \begin{bmatrix} b \\ 0 \end{bmatrix}, \qquad \text{and } c = [c, 0]$$

Considering zeroing for Eq. (5.22):

$$\tilde{c}\tilde{h} = c(h - \breve{b}(c\breve{b})^{-1}ch) = 0 \tag{5.23}$$

For most m:

$$\tilde{c}\tilde{b}_1 = c\breve{A}\breve{b} \neq 0$$

Therefore, by applying state feedback control to v_i, zeroing can be achieved. The estimated state $\hat{x}_i$ is used.

In order to stabilize the nonminimum phase system, $\hat{x}_i$ is also feedback to u_i such that

$$v_i = L\hat{x}_i = -Lx_i - Le_i = -\tilde{L}\tilde{x}_i \tag{5.24}$$

$$u_i = -g\hat{x}_i = -gx_i - ge_i = -\tilde{g}\tilde{x}_i \tag{5.25}$$

where

$$\tilde{L} = [L, L] \quad \text{and} \quad \tilde{g} = [g, g]$$

Eqs. (5.24) and (5.25) being substituted into Eq. (5.22a):

$$\bar{x}_{i+1} = (\tilde{A} - \tilde{b}_1\tilde{L} - \tilde{b}_2\tilde{g})\bar{x}_i + \tilde{h}\,d_i \qquad (5.26)$$

where

$$\tilde{A} - \tilde{b}_1\tilde{L} - \tilde{b}_2\tilde{g} = \begin{bmatrix} A - \tilde{A}\hat{b}L - \check{b}g, & \check{b}(c\check{b})^{-1}ch(c\hat{h})^{-1}c\hat{A} - \check{A}\hat{b}L - \check{b}g \\ 0, & A - h(c\hat{h})^{-1}c\hat{A} - k \end{bmatrix}$$

When v_i in Eq. (5.24) are applied for zeroing in the transfer function between the input v_i and the output y_i, pole-zero cancellation should be performed.

Load Insensitive Control

This zeroing technique is explained in another way: Fig. 5.2 shows a compensator and a servo motor drive system. The compensator is the inverse system of the motor electrical circuits. The external disturbance includes the equivalent external disturbance due to parameter variations (ΔK, ΔK_t, ΔL, and ΔR). The transfer function $T(s)/U(s)$ is infinitive, since

$$\frac{\hat{K}_a\hat{K}_t/(LS + R)}{1 - 1}$$

Therefore, the transfer function is zero. Actually, the inverse system is approximated by introducing a low pass filter $1/(s + K_f)$ (observer), since the compensator contains the derivative operator, which must be avoided in controllers.

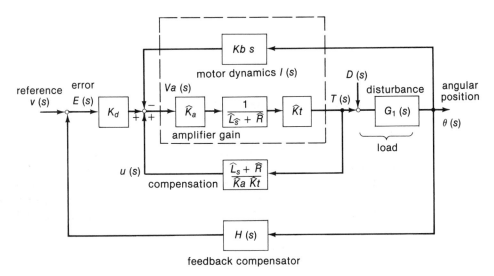

Figure 5.2 Servo control for insensitive to load variations (robust control)

Equivalent Sliding Mode Control*

This is also explained by using sliding mode control theory.

Consider the following linear system represented in the ARMA form:

$$y(k + 1) = A(q^{-1})y(k) + B(q^{-1})u(k) + D(q^{-1})v(k) \qquad (5.27)$$

where q^{-1} is the unit delay operator (z). The polynomials $A(*)$, $B(*)$, and $D(*)$ are of degree n, m, and r, respectively, defined as follows:

$$A(q^{-1}) = a_0 + a_1 q^{-1} + \cdots + a_n q^{-n}$$

$$B(q^{-1}) = b_0 + b_1 q^{-1} + \cdots + b_m q^{-m} \qquad (5.28)$$

$$D(q^{-1}) = d_0 + d_1 q^{-1} + \cdots + d_r q^{-r}$$

where $m < \, = n$ and $r < \, = n$. It is possible to express a_i and b_i as follows:

$$a_i = \hat{a}_i + \delta a_i; \qquad i = 0 \ldots n$$
$$b_i = \hat{b}_i + \delta b_i; \qquad i = 0 \ldots m \qquad (5.29)$$

where a_i and b_i denote the expected values, and a_i and b_i, the uncertainties.

For a trajectory control problem, the sliding surface is defined in the error space as follows:

$$s(k + 1) = C(q^{-1})e(k + 1) = 0 \qquad (5.30)$$

where

$$C(q^{-1}) = 1 + c_0 q^{-1} + \cdots c_{n-1} q^n \qquad (5.31)$$

and

$$e(k) = r(k) - y(k) \qquad (5.32)$$

where $r(k)$ is the reference value.

Equivalent control, which is the term used in sliding mode control, by definition is the control input which makes $s(k + 1)$ equal to zero. If this can be ensured for all k, the system dynamics will clearly be governed by Eq. (5.30) without the influence of any external disturbances and internal uncertainties. If $r(k)$ is a fixed reference value, the error dynamics will arbitrarily be determined by the choice of $C(q^{-1})$. On the other hand, if $r(d)$ is a predetermined trajectory, the error dynamics between the reference trajectory and the output can again be arbitrarily specified. It is obvious that for the stability of the control system, Eq. (5.31) should describe a stable polynomial, i.e., all roots should lie inside the unit circle.

As a first step in the derivation of the control algorithm, let us initially assume that $A(*)$ and $B(*)$ are known polynomials. Neglecting the disturbance terms, the

---

* See references [5.2] and [5.3].

equivalent control, i.e., the control to be applied at the instant k, to make $s(k + 1)$ equal to zero, can easily be obtained from Eqs. (5.27), (5.30), and (5.32) as

$$u_{eq}(k) = -(1/b_0)\{F(q^{-1})y(k) + B_1(q^{-1})u(k) - C(q^{-1})r(k + 1)\} \qquad (5.33)$$

where

$$B_1(q^{-1}) = B(q^{-1}) - b_0$$

$$C_1(q^{-1}) = q\{C(q^{-1}) - 1\}$$

$$F(q^{-1}) = A(q^{-1}) + C_1(q^{-1}) \qquad (5.34)$$

The application of the equivalent control obtained to the system yields the following output at the $(k + 1)^{th}$ instant as

$$y(+ 1) = -C_1(q^{-1})y(k) + D(q^{-1})v(k) \qquad (5.35)$$

In this case, the value assumed by $s(k + 1)$ will not be zero but will be equal to

$$s(k + 1) = D(q^{-1})v(k) \qquad (5.36)$$

It is therefore necessary to modify Eq. (5.33) in an adaptive manner to compensate for the effect of the disturbance terms, as well as the uncertainties in the parameters, which can be considered as additional disturbances. One way of doing this is obvious from Eq. (5.36), since the deviation from the sliding surface is a measure of the disturbance. If the disturbance dynamics are assumed to be slow with respect to the sampling frequency utilized, i.e., if

$$D(q^{-1})v(k) = D(q^{-1})v(k - 1) \qquad (5.37)$$

then the corrective term to be added to Eq. (5.33) can easily be recursively computed. However, the attention in this section is on another approach of a predictive nature, as explained below.

Let us, first of all, assume that the measurement process is not corrupted with noise, i.e., it is deterministic. This means that the uncertainties in the system dynamics arise from the parametric deviations and external disturbances. Therefore, the predicted value of $s(k + 1)$ based on the measurements up to the kth instant can be expressed as

$$s(k + 1 \mid k) = e(k + 1 \mid k) + C_1(q^{-1})e(k) \qquad (5.38)$$

where $\hat{e}(k + 1)$ is the predicted value of the error at $(k + 1)$, given by

$$\hat{e}(k + 1 \mid k) = r(k + 1) - \hat{y}(k + 1 \mid k)$$
$$= r(k + 1) - \hat{A}(q^{-1})y(k) - \hat{B}(q^{-1})u(k) \qquad (5.39)$$

The prediction error that occurs at the kth instant can be written as

$$\tilde{s}(k \mid k + 1) = s(k) - \hat{s}(k/k - 1) \qquad (5.40)$$

where $s(*)$ denotes the actual value of $s(*)$ obtained from the plant output. Now, using Eq. (5.40) as an innovation term, the predicted value of $s(k + 1)$ can be corrected according to the prediction error at the kth instant such that

$$\hat{s}(k + 1 \mid k, k - 1) = \hat{s}(k + 1 \mid k) + \bar{s}(k \mid k - 1) \tag{5.41}$$

The goal now is to find a suitable control law that will render the lefthand side of Eq. (2.31) equal to zero. Since

$$\bar{s}(k \mid k - 1) = \bar{e}(k \mid k - 1) = -\bar{y}(k \mid k - 1) \tag{5.42}$$

Eq. (5.41) can be written as

$$\hat{e}(k + 1 \mid k) + C_1(q^{-1})e(k) - \bar{y}(k \mid k - 1) = 0 \tag{5.43}$$

From Eqs. (5.43) and (5.40), the necessary control is obtained as

$$u(k) = -(1/\hat{b}_0)\{\hat{F}(q^{-1})y(k) + \hat{B}_1(q^{-1})u_k - C(q^{-1})r(k + 1) - y(k \mid k - 1)\} \tag{5.44}$$

If Eq. (5.44) is compared with Eq. (5.33), it is seen that

$$u(k) = u_{eq}(k) - (1/\hat{b}_0)\bar{y}(k \mid k - 1) \tag{5.45}$$

A term proportional to the prediction error at the kth instant is added the equivalent control. The application of Eq. (5.44) will make the system asymptotically stable within the dynamics described by Eq. (5.31). External disturbances will thus be rejected. In the presence of parametric uncertainties, the algorithm allows the use of the knowledge about the uncertainties through the innovation process. Although, for clarity, an SISO system is considered in this section, the extension to multivariable situations is a simple matter.

A control scheme is illustrated in Fig. 5.3. There are two control loops. One is a feedback loop for model matching. The other is a loop for zeroing with an observer that cancels the effect of the equivalent disturbances due to system parameter variations. Thus, a robust model matching is achieved. A flow chart of the control algorithm is given in Fig. 5.4. Presently, the magnitude of the control at the sampling instant is so large that saturation of the input occurs. In order to avoid saturation, the following control scheme is considered and the initial algorithm is slightly modified.

In order to reduce the magnitude of the control, a kind of predictive control is introduced. Let $e(k)$ be the $k'th$ error. Consider a cost function J such that

$$J = \langle \text{sum} \rangle\, 0,\, 3[e(K + j)]^2 \tag{5.46}$$

Assume that $u(k) = u(k + 1) = u(k + 2) = u(k + 3)$. Then, find $u(k)$ (= control) so that J is minimized.

The proposed algorithm is derived on the basis of variable structure control theory. Therefore, this method is explained in terms of sliding mode control. The model used here corresponds to the sliding curve. The model matching control

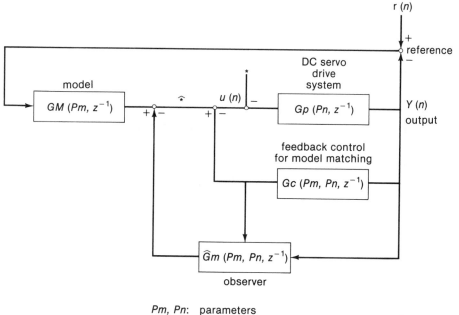

Figure 5.3 Control scheme

is equivalent control in the sliding mode. The equivalent disturbance is estimated by an observer, then it is fed into the input to cancel it (zeroing technique).

The proposed method is believed to be the promising approach among available control algorithms for robust control system, with which accurate motion control in integrated manufacturing systems can be expected. It is noted that power electronics and microelectronics technologies play important roles in realization of the control scheme.

Current and Inverter Control

The following differential equation around the motor armature circuit governs in the time domain:

$$L\dot{i}_a + Ri_a + K_\omega w = KVin + d \qquad (5.47)$$

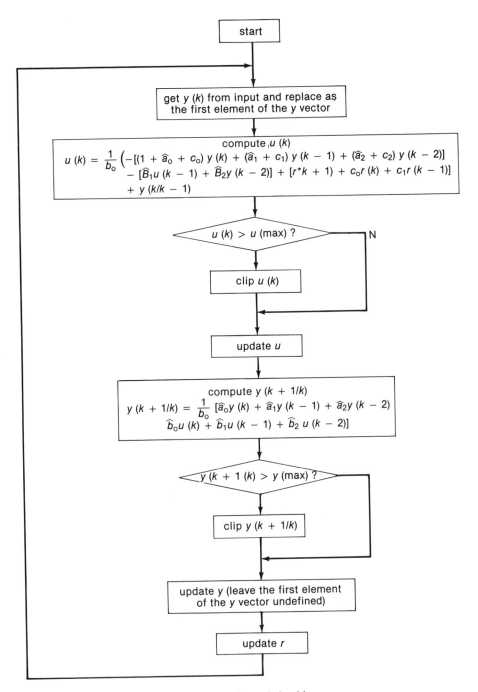

Figure 5.4 Control algorithm

where $d \triangleq$ disturbance
         $ia \triangleq$ motor armature current
         $L \triangleq$ armature inductance
         $R \triangleq$ armature resistance
         $K_\omega \triangleq$ counter electric motive force coefficient
         $K \triangleq$ amplifier gain
         $V_{in} \triangleq$ input
         $w \triangleq$ motor speed

Taking the Laplace Transformation on both sides of Eq. (5.47):

$$s(\hat{L} + \Delta L)i_a(s) + (\hat{R} + \Delta R)i_a(s) + (\hat{K}\omega + \Delta K\omega)\omega(s)$$
$$= (\hat{K} + \Delta K)Vin(s) + d(s) \qquad (5.48)$$

where

$$L = \hat{L} + \Delta L, \quad R = \hat{R} + \Delta R, \quad Kw = \hat{K}w + \Delta Kw \quad \text{and} \quad K + \Delta K$$

$(\hat{\ })$ represents the nominal value and (Δ) shows its deviation (variable part).
   Define the equivalent external disturbance $T_e(s)$ as follows.

$$T_e(s) = s\Delta Li_a(s) + \Delta Ri_a(s) + \Delta K_\omega w(s) - \Delta KVin(s) - d(s)$$

Then Eq. (5.48) becomes

$$(s\hat{L} + \hat{R})i_a(s) + \hat{K}_\omega w(s) + Te(s) = \hat{K}Vin(s) \qquad (5.49)$$

   The block diagram for Eq. (5.49) is shown in Fig. 5.5(a). In order to obtain the estimate $\hat{T}_e(s)$ of $T_e(s)$, the following observer is constructed by using a low pass filter $1/(s + k)$.

$$\hat{T}_e(s) = [\hat{k}V_{in} - \hat{K}_w\omega(s) + (s\hat{L} + \hat{R})i_a(s)]\left(\frac{1}{s + k}\right) \qquad (5.50)$$

It is assumed that $T_e(s)$ changes slowly so that the observer can follow. Next, the estimate $[T_e(s) + \hat{K}_w(s)]$ is fed forward into the input $V_{in}(s)$ so that $T_e(s)$ can be cancelled as shown in Fig. 5.5(b). Then the equivalent block diagram to Fig. 5.5(b) is derived in Fig. 5.5(c). The motor armature circuit can be represented by using the nominal values; therefore, no identification of parameters is required.

   A predictive or a feedforward (dead beat) current controller is designed for the known system shown in Fig. 5.5(c). The proposed fast and accurate current control block diagram is shown in Fig. 5.5(d).

   The voltage equation of the motor is given as follows:

$$\hat{K}_{Vin} = \left(\hat{R} + \frac{d\hat{L}}{dt}\right)i_a \qquad (5.51)$$

By the following approximation

$$\frac{di_a}{dt} = \frac{i_a(n + 1) - i_a(n)}{T} \qquad (5.52)$$

Where T is the sampling time, and n is the sampling number, Eq. (5.51) can be written as

$$\hat{K}V_{in}(n) = \hat{R}i_a(n) + [\hat{L}/T)(i_a(n + 1) - i_a(n)] \tag{5.53}$$

Consider the predictive voltage which makes the current $i_a(n + 1)$ equal to the reference $i_a^*(n + 1)$.

From Eq. (5.53), the predictive voltage is given by the following equation:

$$\hat{K}V_{in}(n) = \hat{R}i_a(n) + (\hat{L}/T)\{i_a^*(n + 1) - i_a(n)\} \tag{5.54}$$

On the other hand, the current $i_a(n)$ is given by Eq. (5.55), replacing n by $n - 1$ in Eq. (5.54).

$$i_a(n) = i_a(n - 1) + (T/\hat{L})\{\hat{K}V(n - 1) - \hat{R}i_a(n - 1)\} \tag{5.55}$$

This equation shows that the current at the nth sampling n can be calculated by the current and voltage prior to the nth sampling.

By substituting Eq. (5.55) into Eq. (5.54), the following equation is obtained:

$$\hat{K}V_{in}^*(n) = 2\hat{R}i_a(n - 1) + (\hat{L}/T)\{i_a^*(n + 1) - i_a(n - 1)\} - \hat{K}V_{in}(n - 1)$$

$$\tag{5.56}$$

The predictive voltage at the nth sampling can be calculated by using the current and the voltage prior to n. Therefore, one sampling is saved as computational time. This is realized with software of the TMS 320 family (DSP).

In the same way, this robust predictive control method is applied to a brushless motor current control, represented by the brushless servo motor model (physical mode) in section 3.2. Then the optimal predictive control voltage $V_{in}^*(n)$ is obtained [5.4].

On the basis of the calculated voltage, the PWM vector selection to control the inverter is performed as follows: $V_{in}^*(n)$ is synthesized by the combination of $V(100)$ and $V(110)$ [5.5] and zero vector as shown in Fig. 5.6. The time interval for each vector is easily calculated and is controlled by the interrupt from the interval timer of DSP. The method is similar to the conventional subharmonic PWM, but the switching frequency would be reduced to $\frac{2}{3}$ when $V_{in}^*(n)$ is within a hexagon shown in Fig. 5.6.

Caution: This current control method has not yet been implemented. This may not work due to high pass filter implementation, [5.4] is recommended. An adaptive scheme is used instead of this zeroing with an equivalent observer; this method is provided only for explanation.

Speed and Force Control

Speed control.   Assume that the previous predictive (or high gain) current control is to be used. Then, the equivalent control block diagram for a mechanical part of a dc servo motor can be drawn as shown in Fig. 5.7(a). The equation is

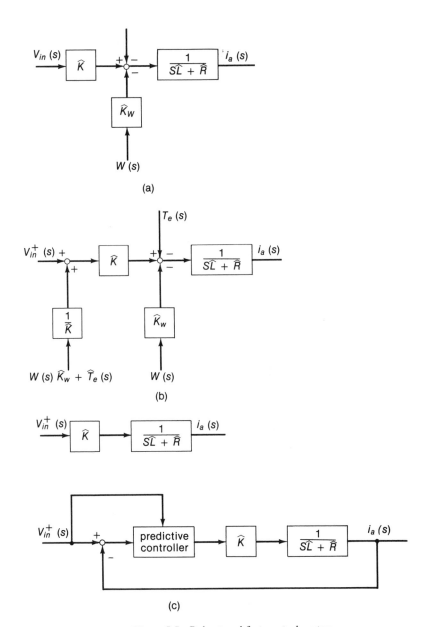

Figure 5.5   Robust and fast control system

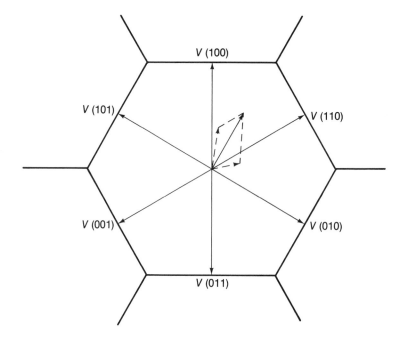

Figure 5.6 Average voltage PWM

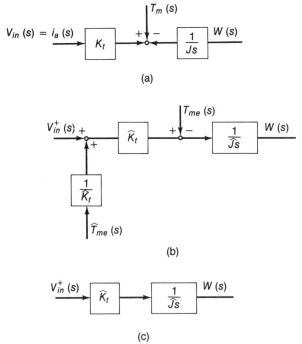

Figure 5.7 Deviation of fast and accurate speed control scheme

as follows:

$$sJw(s) + T_m(s) = K_e i_a(s) \qquad (5.57)$$

where J is the total inertia, K_e is a torque constant, and $T_m(s)$ represents the total disturbance such as frictions, load torques, external forces, and torque disturbances (ripples). Let J be $(\hat{J} + \Delta J)$, and K_t be $(\hat{K}_t + \Delta K_t)$. Then Eq. (5.57) can be rewritten as

$$s(\hat{J} + \Delta J)w(s) + T_m(s) = (\hat{K}_t + \Delta K_t)i_a(s) \qquad (5.58)$$

Define the equivalent external disturbance T_{em} to be

$$T_{em}(s) = s\Delta J_w(s) - \Delta K_t i_a(s)$$

Then Eq. (5.58) becomes

$$s\hat{J}w(s) + T_{em}(s) = \hat{K}_t i_a(s) \qquad (5.59)$$

The estimate $T_{em}(s)$ of $T_{em}(s)$ can be constructed with the following observer by using a low pass filter $1/(s + K_q)$.

$$\hat{T}_{em}(s) = [-s\hat{J}w(s) + \hat{K}_t i_a(s)]/(s + K_q) \qquad (5.60)$$

Then $\hat{T}_{em}(s)(1/\hat{K}_t)$ is fed forward into the input in order to cancel $T_{em}(s)$, as shown in Fig. 5.7(b). The resulting equivalent control block diagram is shown in Fig. 5.7(c).

The servo motor is represented by known and fixed parameters; therefore, robust predictive (feedforward) control is applicable to obtain a fast speed response.

Force control.   In the same way described in Fig. 4.7, force control can be performed easily, since motor parameters are known and fixed. Figure 5.8 shows force control by using zeroing with an equivalent observer.

Another robust control method is high gain control. Figure 5.9 (Example) shows a block diagram of motor speed control with a sliding mode method. It is noted that a current sensor is not used. The next two sections cover robust high gain controllers.

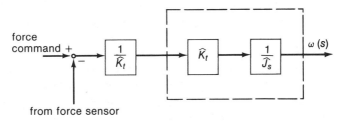

Figure 5.8   Force control after applying zeroing with an equivalent disturbance observer

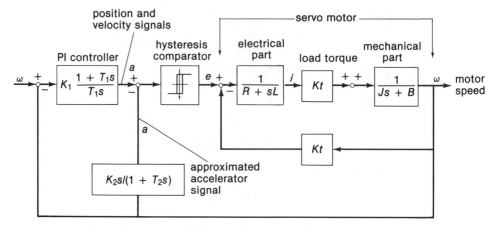

Figure 5.9 Example of sliding motor speed controller without current feedback loop

5.2 IMPROVED SLIDING MODE CONTROL WITH VARIABLE GAIN FEEDFORWARD PATH

Position Tracking Control [5.6]

A variable structure control scheme, which consists of continuous adaptive gain feedback and feedforward controls, with an equivalent external disturbance observer, is developed to achieve accurate (robust) decoupled model, following control for a manipulator powered by PWM transistor converter-fed dc servo motors. The proposed control scheme is implemented by a fast digital signal processor (DSP). It has been confirmed in experiments that the position trajectories are smooth and track the desired trajectories accurately (robustly).

The variable structure control with a sliding mode yields robust control (with which accurate tracking control can be expected); therefore, it has been applied to motion control recently. State of the art in theory and application of discontinuous systems is well explained in other references, so it is not described here.

The main drawback of sliding mode control, however, is that the control action is discontinuous and causes the undesirable chattering. Thus, in order to cope with this, a novel variable structure control method is proposed.

Accurate model following (tracking of desired trajectories) is the control challenge in the development of modern industrial robots and manipulators in a flexible manufacturing system environment. There are well-known nonlinearities in the overall system. Several methods are proposed to overcome this. Nonlinear compensations are devised. However, they are complex and costly to implement, since such schemes suffer from the requirement of a detailed model of the manipulator and load forecasting. They cannot be implemented. Model reference adaptive control (MRAC) is also applied to motion control of a manipulator. However, the global stability of MRAC system in the case of uncertain nonlinear plants is problematic, especially since the MRAC is sensitive to external disturbances and yields very slow dynamic characteristics. The sliding mode control gives robust control (with which we can expect accurate tracking). However, the main drawback of this approach is that the control action is discontinuous and results in causing the undesirable chattering phenomenon.

This section presents a methodology of a modified model following control with an improved sliding mode control. The proposed method is based on a variable structure control scheme, which consists of continuous adaptive gain feedback, feedforward controls, and a forced model. The feedforward path is added in order to cancel external disturbances roughly and to make time responses faster. The feedback loop is for the purpose of obtaining the strong convergence properties of the error to the origin. That the proposed improved sliding mode control is continuous not only guarantees that the error remains bounded, but also that it tends to an arbitrarily small neighborhood of the origin.

In addition, an equivalent external disturbance observer, with which the effects of external disturbances, including the disturbance due to system parameter variations are roughly canceled, is constructed since (in general) the persistent external disturbances cause the undesirable chattering in the sliding mode.

The proposed method is practically applied to a decoupled model following control for a manipulator. The control strategies are implemented by an available DSP (TI-TMS 320C25). This DSP plays an important role in increasing the speed of motion.

The robustness of the obtained controller, which results in accurate decoupled tracking, is confirmed by experiments. It is found that the controller is easily designed since it requires less knowledge of the plant.

The overall control scheme is shown in Fig. 5.10. The goal is to find adaptive gain feedback and feedforward controls to nullify the error $e(t)$ as $t \rightarrow \infty$. In order to make time response faster and to cancel the external disturbance roughly, the next adaptive feedforward control u_{ffi} is applied.

$$u_{ffi} = K_{fDi} \, \phi D_i \, | \, \dot{g}_i \, |, \qquad \phi D_i = 1 - \frac{Sl_i}{| \, Sl_i \, | + \delta_1} \frac{\dot{g}_i}{| \, \dot{g}_i \, |} \qquad (5.61)$$

$$Sl_i = C_i e_{i1} + e_{i2} = b_{0i}^T P_i [e_{i1}, e_{i2}]^T = 0 \qquad (5.62)$$

where δ_i is a positive constant and p_i is a positive definite matrix. The output tracks g_i roughly only with u_{ffi} applied. K_{fDi} is roughly determined by experiment such that

$$0 = f_i(g, \dot{g}) + b_i(g) K_{fDi} | \, \dot{g}_i \, | - \ddot{g}_i \qquad (5.63)$$

Next, an adaptive feedback gain control u_{fbi} is applied in order to stabilize the overall system asymptotically.

$$u_{fbi} = K_{Pi} k_{vi} | \, e_{i1} \, | \qquad (5.64)$$

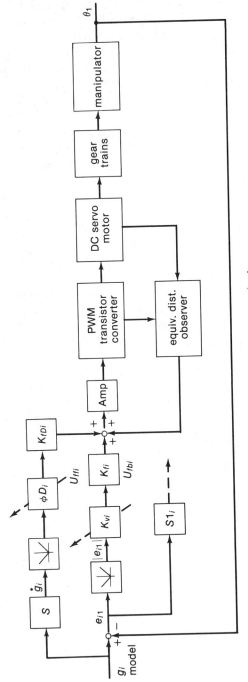

Figure 5.10 Overall control scheme

122

where

$$K_{vi} = \frac{-Sl_i}{|Sl_i| + \delta_i}$$

and δ_i is a positive constant.

The design procedure for the controller is as follows:

1. Determine a Liapunov function $v_i(e_i)$, then evaluate a sliding curve

$$Sl_i = b_{0i}^T P_i [e_{i1}, e_{i2}]^T \qquad (5.65)$$

2. Calculate the controls, u_{ffi} and u_{fbi}. The convergence of the error between the outputs of the model and control object to the origin is proved.

As a modern robot manipulator is required to have quick and precise response, the fast servo system and low ratio gear (sometimes a direct drive system) tend to be adopted in each joint motion control system. In such a case, coupled force disturbances among the joints become significant.

In general, sliding mode control causes the undesirable chattering when persistent external disturbances are applied. Therefore, a robust and fast torque observer, whose derivation is based on zeroing techniques with an equivalent observer described in section 5.1, is constructed.

The total sum of coupled inertial forces, centrifugal forces, coriolis forces, friction, viscosities, payloads, gravity, stray forces, and even the torques due to system parameter variations, can be estimated to be T with nominal observer parameters used.

Then, the estimated equivalent external disturbance torque, T, is fed into the control input to cancel the effect of the disturbances. The bandwidth of the filter should be wider than that of the external signal T; otherwise, stability problems arise. If the disturbance observer is used, then the adaptive feedforward controller can be replaced by an inverse system of the disturbance-cancelled system.

5.3 VARIABLE STRUCTURE PI WITH VARIABLE GAIN FEEDFORWARD, INPUT MODULATION AND EQUIVALENT DISTURBANCE OBSERVER*

Position Tracking Control

A new variable structure control scheme, which consists of continuous adaptive gain geedback (PI) and feedforward controls, is developed to achieve an accurate decoupled model, following in a class of nonlinear time-varying systems in the

_____
* See reference [5.7].

presence of disturbances, parameter variations, and nonlinear dynamic interactions. The developed method is then practically applied to decoupled model following motion control, for a two-degrees-of-freedom manipulator, powered by PWM transistor converter-fed servo motors. The overall control strategies are implanted with a DSP (TI, TMS320C25). It is found through the experiments that this controller is simple, easily designed, and performs quite satisfactorily.

This section presents a methodology of modified model following control, with an improved sliding mode control. The proposed method is based on a variable structure control scheme, which consists of continuous adaptive gain feedback (PI), feedforward controls, and a forced model. The feedforward path is added in order to cancel external disturbances roughly and to make time responses faster. The feedback loop is for the purpose of obtaining the strong convergence properties of the error to the origin. The proposed improved sliding mode control is continuous. That not only guarantees that the error remains bounded, but also that it tends to an arbitrarily small neighborhood of the origin.

Then the proposed method is practically applied to decoupled model following control for a two-degrees-of-freedom manipulator. The control strategies are implemented by an available DSP (TI, TMS320C25). This DSP plays an important role in increasing the speed of motion.

The robustness of the obtained controller, which results in accurate decoupled tracking, is confirmed by the experiments. It is found that the controller is easily designed since it requires less knowledge of the plant.

The overall control scheme is shown in Fig. 5.11. The goal is then to find adaptive gain feedback and feedforward controls to nullify the error $e(t)$ as $t \to \infty$. In order to cancel the external disturbances roughly and to make time response faster, u_{ffi}, the adaptive feedforward control, is applied as described in Eq. (5.61). This acts as a derivative controller. Then, an adaptive gain feedback control (PI) v_{fb} is applied in order to stabilize the overall system asymptotically.

$$u_{fbi} = K_{vpi} \mid e_1 \mid + K_{vli} \int \mid e_i \mid dt \tag{5.66}$$

$$K_{vpi} = \frac{Sl_i}{\mid Sl_i \mid + \delta_{bi}} K_{pi} \tag{5.67}$$

$$K_{vil} = \left(1 - \frac{\mid Sl_i \mid}{\mid Sl_i \mid + \delta_{bi}} \right) \frac{Sl_i}{\mid Sl_i \mid} K_{il} \tag{5.68}$$

where b_i, K_{pi}, and K_{li} are positive constants. Figure 5.12 shows the relation of K_{vpi} and K_{vli} versus sl_i. Near the sliding curve, an integral controller dominates.

The design procedure for the controller is as follows:

1. Determine a Liapnov function $V_i(e_i)$, then evaluate a sliding curve, $Sl_i = b_{0i}T_{pi}[e_{i1}, e_{i2}]$ equation (5.65).
2. Calculate the controls, U_{ffi} and U_{fbi}. The convergence of the error between the outputs of the model and the control object to the origin is proved.

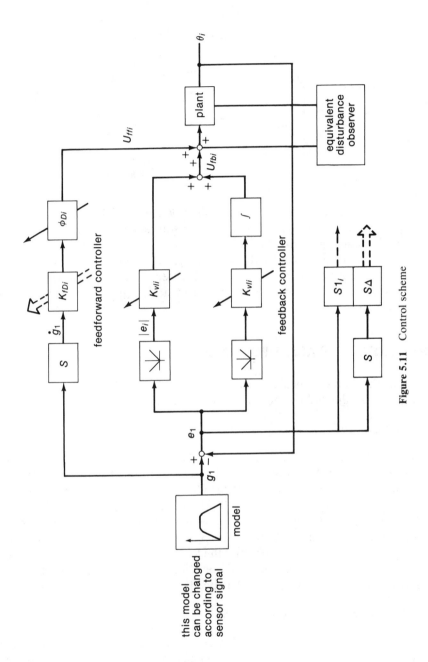

Figure 5.11 Control scheme

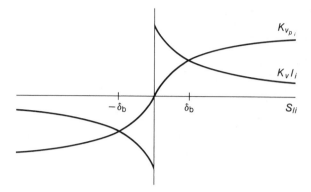

Figure 5.12 K_{vpi} and K_{vli} versus S_{li}

If necessary (for example, the disturbance has a large magnitude and high frequency components), an equivalent disturbance observer may be added to cancel the equivalent disturbance.

The input command may be changed according to the sensor signal (for example, identified inertia, etc.). This is effective to increase control capability (for example, anti wind-up phenomena of an integrator controller can be prevented).

5.4 MODEL REFERENCE ADAPTIVE CONTROL

The application of this algorithm is limited to the system which has two or three parameters adjusted by the following adaptive mechanism.

Speed Control*

Model reference adaptive control (MRAC) is applied to microprocessor-based, adjustable-speed dc motor drives. The algorithm of the MRAC is based on the linear model following control (LMFC) and is the combination of the adaptive controller with the LMFC. The MRAC-based speed controller allows the indistinctness and/or inaccuracy in the motor and load parameters in the system design stage. It also maintains the prescribed control performance in the presence of the motor parameter perturbations and the load disturbances. The experimental setup is constructed using a microprocessor, and the experimental results confirm the useful effects of the MRAC-based speed controller.

MRAC Algorithm Based on LMFC

Preparation.   The motor drive system is generally a second or higher order system; it has at least two dynamics in electrical and mechanical responses. The

* See reference [5.8].

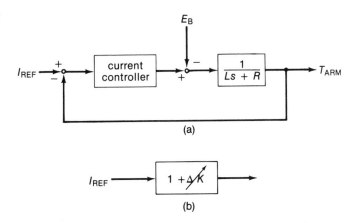

(a)

(b)

Figure 5.13   Effect of current controller. (a) Current control minor loop. (b) Equivalent block.

first order MRAC algorithm is, however, realistic for the capability of the microprocessors available on the market at present.

The MRAC-based speed controller has a current controller as its minor loop. The control logic of the current controller is proportional-integral (PI) and the P- and I-gains are determined, as to make the rise time of the current response shorter than that obtained by the conventional PI-controller design practice. The shorter rise time means the faster termination of the transient at every control period. The currrent minor loop, regulated as mentioned above, can be regarded as a variable-gain component with a value that fluctuates around unity as illustrated in Fig. 5.13; if thus regulated, current response is much faster than the prescribed speed response. The current minor loop has the effect of eliminating the influence of the back electromotive force (E) as well.

The design technique reduces the transfer function between the current reference (I) and the rotor speed (N) to the first order, as shown in Fig. 5.14. The first-order MRAC thus becomes applicable to the dc motor speed control.

Derivation of MRAC algorithm.   The MRAC algorithm is based on LMFC and can be referred to as adaptive model following control (AMFC) as well.

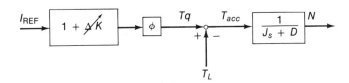

Figure 5.14   Approximated electromechanical dynamics

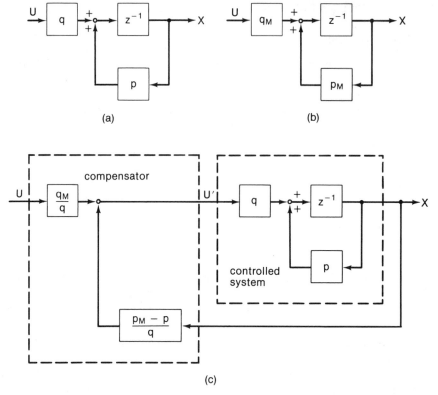

Figure 5.15   Derivation of typical LMFC. (a) Controlled system, (b) reference model, and (c) model-following system.

Figure 5.15 explains the construction of a typical LMFC on which the MRAC algorithm is based. Figures 5.15(a) and 5.15(b) are the block diagrams of the first-order controller system and the first-order reference model in time-discrete form, respectively. The p, q, p_M, and q_M are defined as follows:

$$p = \exp(-T_c/T_s) \qquad (5.69)$$

$$q = K(1 - p)(K\text{: gain}) \qquad (5.70)$$

$$p_M = \exp(-T_{cM}/T_s) \qquad (5.71)$$

$$q_M = K_M(1 - p_M)(K_M\text{: gain}) \qquad (5.72)$$

where T_c and T_{cM} are the time constants of the controlled system and the reference model, respectively, and the T is the sampling period. The objective of LMFC is to make the response of the controlled system follow that of the reference model. The compensation gains placed in front of the controlled system, as in Fig. 5.15(c), serve the above objective. A closer look at Fig. 5.15(a) reveals that it is identical to Fig. 5.15(b). The identicalness between them guarantees the model-following capability of the LMFC.

In the application of the adaptive control system, the controlled system parameters p and q are supposed to be unknown and/or variable, and they are denoted by $p(k)$ and $q(k)$, as in Fig. 5.16(a). Accordingly, the compensation gains can no longer be constant. They should be adjusted during operation through the estimation of $p(k)$ and $q(k)$:

$$G(k) = \frac{q_M}{q(k)} \tag{5.73}$$

$$G(k) = \frac{p_M - p(k)}{q(k)} \tag{5.74}$$

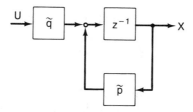

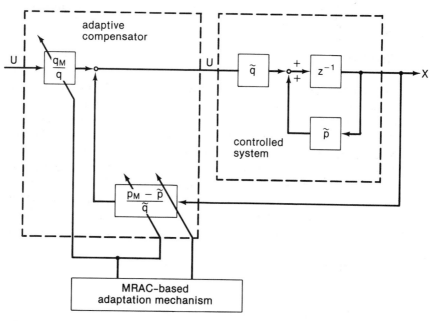

Figure 5.16   Fundamental structure of MRAC-based control system. (a) Controlled system. (b) Adaptive model-following system.

The adaptation mechanism takes care of the estimation as illustrated in Fig. 5.16(b). The correct and quick estimation of $p(k)$ and $q(k)$ can reduce Fig. 5.16(b) to Fig. 5.15(b) and realize the following adaptive model.

The estimation algorithm is based on the general theory of MRAC and is given as:

$$p(k + 1) = p(k) + K_p x(k) e^*(k + 1) \tag{5.75}$$

$$q(k + 1) = q(k) + K_q U'(k) e^*(k + 1) \tag{5.76}$$

where $e^*(k)$ is the adaptive error signal and is defined as

$$e^*(k + 1) = \frac{p_M e^*(k) + x(k + 1) - [p(k)x(k) + q(k)U'(k)]}{1 + K_p x^2(k) + K_q U'^2(k)}$$

$$\tag{5.77}$$

The input reference signal to the controlled system $U(k)$ is then generated as

$$U(k) = \frac{[p_M - p(k + 1) \mid x(k + 1) + q_M U(k + 1)]}{q(k + 1)} \tag{5.78}$$

MRAC-based speed controller.   The basic structure of the MRAC system presented in Fig. 5.16(b) is directly applied to the dc-motor speed-control system as shown in Fig. 5.17. The U and U' in Fig. 5.16(b) correspond to U_M and I_{REF} in Fig. 5.17, respectively. The precompensator is placed in front of the U_M for higher control performance.

Flux Vector Control [5.9]

This section deals with a compensation method of the rotor-resistance variation in induction motor drives using high-performance slip-frequency control. The proposed method is based on a discrete-type model reference adaptive system (MRAS), and it is implemented in an 8086 microprocessor. When an induction motor is driven by a controlled current source, the system sensitivity to the rotor resistance variation is increased. In the MRAS, the value of the rotor resistance is estimated and the slip-frequency gain is adjusted. Experimental and numerical results show that even if the value of the rotor resistance varies from its nominal value, the secondary flux level is maintained constant by using this compensation method. These results point out the validity of the proposed method.

Nomenclature

$e_1 \triangleq$ primary voltage

$e_2 \triangleq$ secondary voltage

$i_1 \triangleq$ primary current

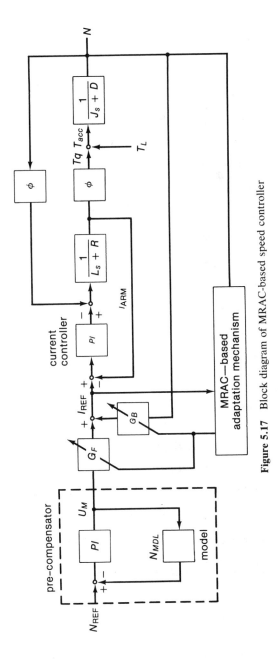

Figure 5.17 Block diagram of MRAC-based speed controller

$i_2 \triangleq$ secondary current

$M \triangleq$ magnetizing inductance

$L_1 \triangleq$ primary inductance

$L_2 \triangleq$ secondary inductance

$l_1 \triangleq$ primary leakage inductance

$l_2 \triangleq$ secondary leakage inductance

$L_\sigma \triangleq$ equivalent total leakage inductance $(= L_1 - M^2/L_2)$

$r_1 \triangleq$ primary resistance

$r_2 \triangleq$ secondary resistance

$\omega_0 \triangleq$ supply frequency

$\omega_r \triangleq$ rotor frequency

$\omega_{\text{slip}} \triangleq$ slip frequency

$T \triangleq$ electrical torque

$T_1 \triangleq$ load torque

$\lambda_2 \triangleq$ secondary flux

$J \triangleq$ total inertia

$D \triangleq$ viscosity resistance

ref $\triangleq$ This upper suffix denotes the reference value of the nominal value.

$\alpha \triangleq$ This lower suffix denotes the -axis component.

$\beta \triangleq$ This lower suffix denotes the -zxis component.

$P \triangleq$ differential operator or Laplace operator

$Z^{-1} \triangleq$ one-step delay operator

High-performance slip-frequency control. The transient state of the induction motor can be analyzed mathematically by using the two-axis theory. To simplify the analysis, the orthogonal coordinates called α–β axes are introduced. The coordinates rotate with the power source frequency ω_0 around the center axis of the motor. In these coordinates, all the variables contain only the *dc* component. The induction motor is represented by the following differential equations:

$$e_{1\alpha} = (r_1 + L_1P)i_{1\alpha} - \omega_0L_1i_{1\beta} + MPi_{2\alpha} - \omega_0Mi_{2\beta} \tag{5.79}$$

$$e_{1\beta} = \omega_0L_1i_{1\alpha} + (r_1 + L_1P)i_{1\beta} + \omega_0Mi_{2\alpha} + MPi_{2\beta} \tag{5.80}$$

$$e_{2\alpha} = MPi_{1\alpha} - (\omega_0 - \omega_r)Mi_{1\beta} + (r_2 + L_2P)i_{2\alpha} - (\omega_0 - \omega_r)L_2i_{2\beta} \tag{5.81}$$

$$e_{2\beta} = (\omega_0 - \omega_r)Mi_{1\alpha} + MPi_{1\beta} + (\omega_0 - \omega_r)L_2i_{2\alpha} + (r_2 + L_2P)i_{2\beta} \tag{5.82}$$

For the caged motor with the rotor circuit shorted, we have

$$e_{2\alpha} = 0 \tag{5.83}$$

$$e_{2\beta} = 0 \tag{5.84}$$

The motor torque is given as follows:

$$T = M(i_{1\beta}i_{2\alpha} - i_{1\alpha}i_{2\beta}) = \lambda_{2\beta}i_{2\alpha} - \lambda_{2\alpha}i_{2\beta} \tag{5.85}$$

Here, λ_2, which represents the secondary flux level, is defined as:

$$\lambda_{2\alpha} = Mi_{1\alpha} + L_2i_{2\alpha} \tag{5.86}$$

$$\lambda_{2\beta} = Mi_{1\beta} + L_2i_{2\beta} \tag{5.87}$$

The high-performance slip-frequency control is used in two ways, depending on the type of power source: a controlled-voltage source and a controlled-current source. Fig. 5.18 shows the block diagram of the induction motor, driven by a controlled-voltage source. In case a controlled current source is used, it is sufficient to consider the area within the broken line in Fig. 5.18.

In the high-performance slip-frequency control, the secondary flux and the torque current must be decoupled by using the slip frequency expressed in Eq. (5.88) and the exciting current command expressed in Eq. (5.89):

$$\omega_{\text{slip}} = \frac{r_2}{L_2i_{1\alpha}} i_{1\beta} \tag{5.88}$$

$$i = \text{constant} \tag{5.89}$$

Equations (5.88) and (5.89) lead to cancellation of the cross terms of the secondary circuit. As a result, both $i_{2\alpha}$ and $\lambda_{2\beta}$ become 0. In the case of the controlled-voltage source, the following equations, as well as Eqs. (5.88) and (5.89), are used to decouple variables of primary circuit:

$$e_{1\alpha} = r_1i_{1\alpha}^{\text{ref}} - \omega_0L_\sigma i_{1\beta}^{\text{ref}} \tag{5.90}$$

$$e_{1\beta} = r_1i_{1\beta}^{\text{ref}} + \omega_0L_1i_{1\alpha}^{\text{ref}} \tag{5.91}$$

$i_{1\beta}^{\text{ref}}$ in Eqs. (5.90) and (5.91) is obtained by subtracting the actual rotor velocity from the reference velocity at the velocity controller. As a summary, the high-performance slip-frequency control is obtained by using Eqs. (5.88) and (5.89) in the case of the controlled-current sources, and Eqs. (5.88) through (5.91) in the case of controlled-voltage sources. By the control using Eqs. (5.88) through (5.91), the torques becomes proportional to the torque reference (or $i_{1\beta}^{\text{ref}}$), and a quick response to the change of torque command can be achieved.

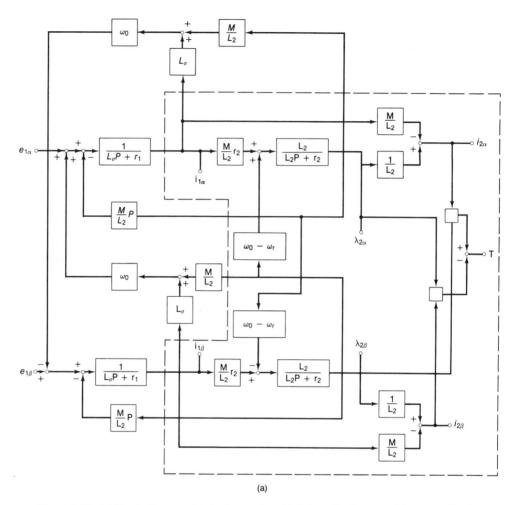

(a)

Figure 5.18 (a) Block diagram of induction motor. (b) Schematic diagram of the controller in case of using controlled-current source. (c) Schematic diagram of the controller in case of using controlled-voltage source.

MRAS for the rotor resistance compensation. Figure 5.19 shows the basic structure of the MRAS, in which the ideal motor runs in parallel with the actual motor. The output of the model is compared with real output. The difference of these outputs (i.e., error) may depend on the parameter variation of the actual motor. If an integral action is employed in the adaptive mechanism, a steady-state offset is eliminated.

The motor model should be determined by applying MRAS to the vector control of the induction motor. Since the time constant for the variation of the rotor resistance is far larger than the time constant of the induction motor and

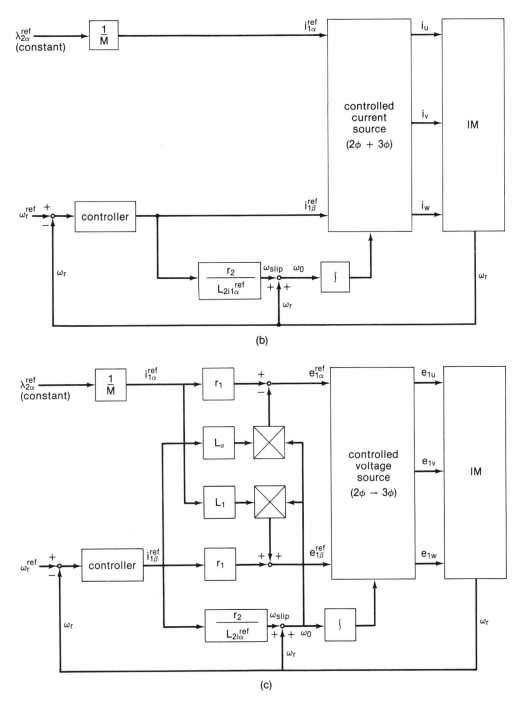

(b)

(c)

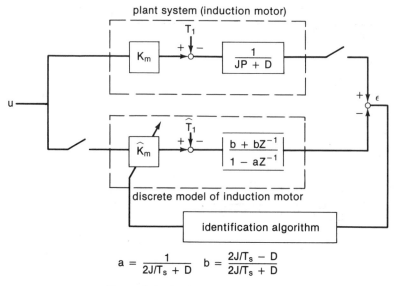

$$a = \frac{1}{2J/T_s + D} \qquad b = \frac{2J/T_s - D}{2J/T_s + D}$$

Figure 5.19   Basic structure of MRAS

the sampling time period, the steady-state model of the induction motor can be adopted. When the vector control algorithm is applied to Fig. 5.13, the total block diagram is represented in Fig. 5.19. Here, T_s is sampling time and K_m may be proportional to the flux level as follows:

$$K_m = \frac{3r_2^{\text{ref}}}{L_2 r_2 i_{1\alpha}} (\lambda_{2\alpha}^2 + \lambda_{2\beta}^2) \tag{5.92}$$

For the compensation of rotor resistance variation, the value of K_m in the motor model should be successively estimated. When the value of K_m is estimated, the varied value of the rotor resistance $\Delta r_2 (= r_2 - r_2^{\text{ref}})$ can be obtained as follows:

$$\Delta r_2 = \frac{A \pm \sqrt{A^2 - 4(k - 1)(k - 3)r_2^{\text{ref}2}i_{1\alpha}^2 + i_{1\beta}^2)}}{2(3 - k)i_{1\alpha}^2}$$

$$(+ \text{ for } i_{1\alpha} > i_{1\beta}, \; - \text{ for } i_{1\alpha} < i_{1\beta}) \tag{5.93}$$

where

$$A = \{2k(i_{1\alpha}^2 + i_{1\beta}^2) - 3i_{1\alpha}^2 - i_{1\beta}^2\}r_2^{\text{ref}}$$

$$K = \frac{3M^2 i_{1\alpha}}{K_m l_2}$$

Parallel structured MRAS estimates K_m successively as follows:

$$K_m(k) = \frac{1}{b} \left\{ b\hat{k}_m(k - 1) + \frac{f\phi}{1 + \phi^T F\phi} \epsilon^0 \right\} \tag{5.94}$$

where F is adaptive gain matrix, of which the matrix manipulation causes the most computation estimating the k_m, f is the second row vector in F. $\phi(k)$ is defined as follows:

$$\phi^T(k - 1) = [y(K - 1), u(k), u(k - 1)] \qquad (5.95)$$

The block diagram of the MRAS, including a block of integral action, is shown in Fig. 5.20 where the input is the torque current command. This command, proportional to the slip-frequency command, is applied to the actual motor and the model motor. On the other hand, the value of load torque is calculated by using the value of K_m. This method brings about the error between the real and the calculated load torques. This difference converges to 0 if the torque-current command contains *ac* components of noise, which are naturally generated by PWM switching action in power converters.

Brushless servo motor current control by MRAC can be found in reference [5.4].

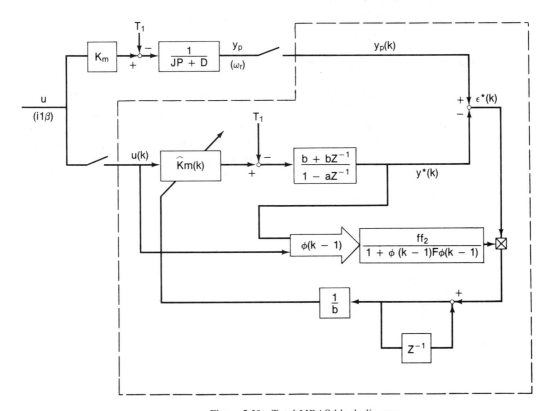

Figure 5.20   Total MRAS block diagram

5.5 LEARNING (REPETITIVE) CONTROL

The repetitive control is effective if the external signals (external disturbances and references) are periodical. The control algorithm is explained in Fig. 5.21. Let

$$U(n) = [u(1), u(2), \ldots, u(N)]^T$$
$$U(n + 1) = [u(N + 1), u(N + 2), \ldots, u(2N)]^T \tag{5.96}$$

and the error

$$E(n) = [e(1), e(2), \ldots, e(N)]^T$$
$$e(i) = r(i) - y(i)$$

where $r(i)$ is the reference and $y(i)$ is the output. Then the repetitive control algorithm is

$$U(n + 1) = U(n) + K[E(n) - E(n - 1)] \tag{5.97}$$

where K is a constant gain matrix.

However, this is not robust. Compensators such as a stabilizer and adaptive method are necessary. Therefore, this method may be combined with the zeroing, with an equivalent observer described in Eq. (5.1).

Current and Inverter Control [5.10]

This section presents a new control method of the switching devices in the inverter system. The simulation results reveal that the output wave form is close to a sinusoidal input waveform, for its fundamental part is corresponding to the sinusoidal input and its harmonics parts are completely eliminated. Furthermore, the influences of disturbance can be eliminated.

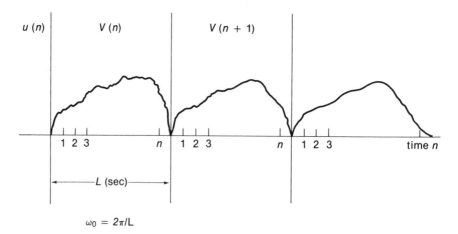

$\omega_0 = 2\pi/L$

Figure 5.21   Learning control algorithm

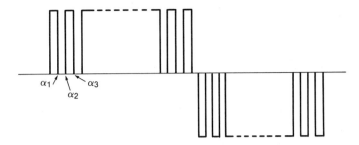

Figure 5.22 Inverter waveform

In the waveform shown in Fig. 5.22, its fundamental component and harmonic components are calculated by the switching timing (α_1, α_2, α_3, . . . , α_n) as follows:

$$Hn = 4/nn \left(1 + \sum_{i=1}^{n} (-1)^i \cos n\alpha_i \right) \tag{5.98}$$

where $n = 1, 3, 5, . . .$ and $Hn = n^{\text{th}}$ harmonic component.

In the previous inverter control (α_1, α_2, . . . , α_n) is calculated with "on-line" and switched so that Hn becomes the desired value. This is feedforward control. However, feedback control (repetitive control) is applied here. Figure 5.23 shows this inverter system.

Error Signal Analyzer

In this section, the error $e(t)$ (the difference between the input and the output) is analyzed and the error vector is derived.

One period (from $e(t - T)$ to $e(t)$, where T is a period) is considered. The following Eq. (5.99) is calculated:

$$a_n(t) = \frac{2}{T} \int_{t-T}^{t} e(\iota) \cos n\omega_0(\iota - t) \, d\iota$$

$$\tag{5.99}$$

$$b_n(t) = \frac{2}{T} \int_{t-T}^{t} e(\iota) \sin n\omega_0(\iota - t) \, d\iota$$

where $n = 1, 2, 3, 4, . . .$

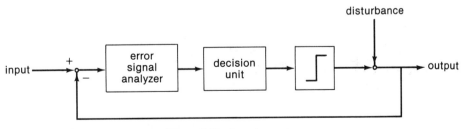

Figure 5.23 Inverter system

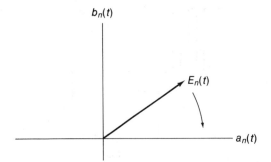

Figure 5.24   Error vector

The error vector $E_n(t)$ is defined by using Eq. (5.99) as follows:

$$E_n(t) = a_n(t) + jb_n(t) \tag{5.100}$$

$E_n(t)$ is explained in the complex plane as shown in Fig. 5.24. It rotates as time elapses.

Decision Unit

In this section, a switching sequence is generated on the basis of the error vector as shown in Fig. 5.25. This decision unit is divided into three blocks.

Block (1) has eight regions, as shown in Fig. 5.26(a), divided by the curve

$$x - y \tan (h\sqrt{x^2 + y^2}) = 0 \tag{5.101}$$

according to the rotating error vector position, the signal shown in Fig. 5.26(b) is generated, Eq. (5.101) is provided to obtain the pulse width T proportional to the magnitude of the error vector $|E_n(t)|$.

In Block (2), the error signal is added to the previous signal and the control signal is generated as shown in Fig. 5.27.

Block (3) constitutes a repetitive controller.

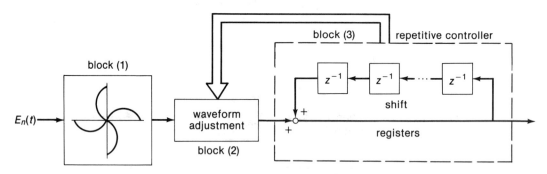

Figure 5.25   Decision unit

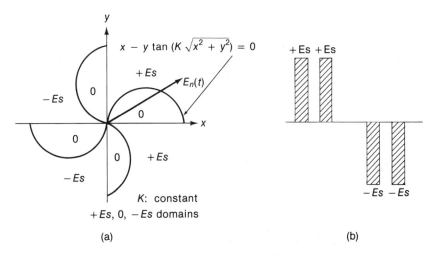

Figure 5.26   (a) Vector plane and (b) error signal

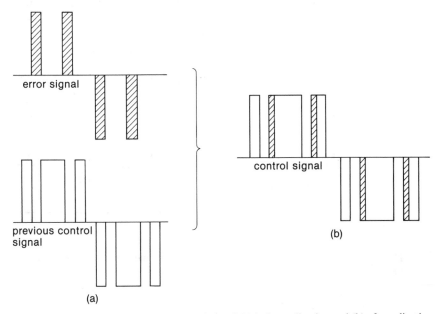

Figure 5.27   Waveform of $e(t)$ and control signal (a) before adjusting and (b) after adjusting

Speed Control*

The repetitive control algorithm is applied to reduce the periodic motor speed fluctuation. The block diagram for this control system is shown in Fig. 5.28.

The transfer function of a repetitive controller at nw_0 is

$$\frac{1}{1 - e^{-jw_0L}} = \frac{1}{1 - e^{-j(2n/L)-L}} = \frac{1}{1 - 1} = \infty \qquad (5.102)$$

where n is an integer ($n = 0, 1, 2, \ldots$). Therefore, the transfer function $w(s)/D(s)$ is zero at nw_0. This is considered as a technique for zeroing at nw_0.

5.6 INTELLIGENT (EXPERT) CONTROL

Fuzzy Expert Grasping Force Control**

There has been significant impact of artificial intelligence on intelligent motion control solutions, as witnessed by the proliferation of knowledge-based expert systems. Fuzzy set control is often used to implement such intelligent controllers. This technique is applied to uncertain systems whose models cannot be utilized to generate robust control. The fuzzy expert system can also make multistage decisions as a general expert makes.

This part presents a fuzzy set controller successfully applied to force control for servo system. The control objective is to grasp balls of various compliances,

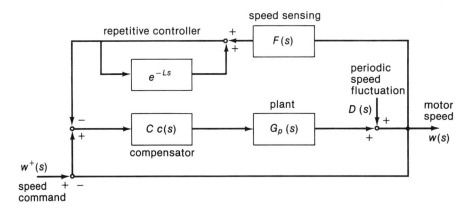

Figure 5.28 Transfer gain between disturbance input and motor speed is close to zero. Lead compensated PLL stable system is used for $Cc(s)$.

* See reference [5.11].
** See reference [5.13].

from a soft tennis ball to a hard steel ball, with constant force command, and achieve uniform dynamic response with no overshoot.

An overall control block diagram is shown in Fig. 5.29. It is known by experience that the control gain K_f is the reciprocal to the compliance K_e. This K_e is determined by using the experiment expressed by a set of fuzzy membership functions F_1 and F_2.

First, K_e is obtained as follows:

$$K_e(k) = [F(k) - F(k - 1)]/V(k - j)$$

Since

$$K_e = dF/dx = (dF/dt)/(dx/dt) = \Delta F/V \qquad (5.103)$$

where $F \triangleq$ measured force, $V \triangleq$ speed command, and $J \triangleq$ servo system delay time.

Then, membership functions F_1 and F_2 are determined by experience. F_1 is for a soft tennis ball and F_2 is for a hard steel ball. They are shown in Fig. 5.30.

$$W_{max} = F_1(K_e) \qquad (5.104)$$

$$W_{min} = F_2(K_e) \qquad (5.105)$$

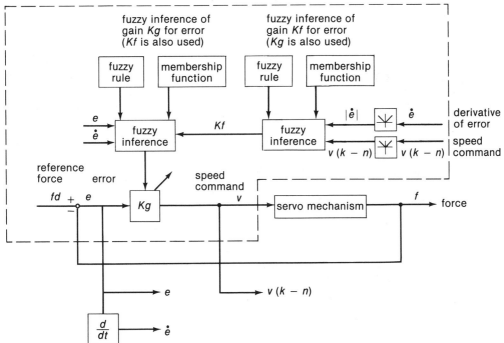

Figure 5.29   Fuzzy set controller

If *ke* is *BG* then *kf* is *kf*max
If *ke* is *SL* then *kf* is *kf*min

(a)

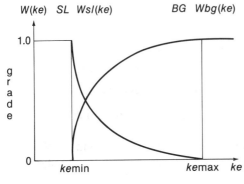

$$Wsl(ke) = \begin{cases} 1 & (0 < ke < kemin) \\ 1.2 - 0.03/ke & (kemin < ke < kemax) \\ 0 & (kemax < ke) \end{cases}$$

$$Wbg(ke) = \begin{cases} 0 & (0 < ke < kemin) \\ 0.03/ke - 0.2 & (kemin < ke < kemax) \\ 1 & (kemax < ke) \end{cases}$$

Figure 5.30 (a) Fuzzy rules and (b) membership function.

(b)

$K_{e\max}(0.2)$ for a hard steel ball and $K_{e\min}(0.02)$ for a soft tennis ball are obtained by experiment.

Finally, decision making (obtaining K_e) is achieved by calculating the following equation:

$$\hat{K}_e = (K_{e\max} W_{\max} + K_{e\min} W_{\min})/(W_{\max} + W_{\min}) \qquad (5.106)$$

This part was on the fuzzy inference of K_e by measuring the force and the speed command [5.12].

The following part deals with the fuzzy inference of the control gain K_g by using the obtained K_t with the error and its derivative [5.13].

The utilized fuzzy rules and membership functions are shown in Fig. 5.31. The final optimal gain K_g is calculated as follows using Fig. 5.31.

Let the membership functions for e and Δe be $u_i(e)$ and $u_j(\Delta e)$, and $w_{ij}(e, \Delta e)_L$ be

$$W_{ij}(e, \Delta e)_L = \mu_i(e) n \mu_j(\Delta e) \qquad (5.107)$$

$$K_g = \frac{\sum W_{ij}(e, \Delta e)_L K_g(e, \Delta e)_L}{\sum W_{ij}(e, \Delta e)_L} \qquad (5.108)$$

	error e							
	−LG	−MD	−MS	−ZO	+ZO	+MS	+MD	+LG
+LG	PM	PM	PM	PM	NM	NV	ZO	PS
+MD	PM	PM	PM	PM	NV	PV	PS	PS
+MS	PM	PM	PM	PM	PV	PS	PM	PM
+ZO	PM	PM	PM	PM	PB	PM	PM	PM
−ZO	PM	PM	PM	PM	PB	PM	PM	PM
−MS	PM	PM	PS	PV	PM	PM	PM	PM
−MD	PS	PS	PV	NV	PM	PM	PM	PM
−LG	PS	ZO	NV	NM	PM	PM	PM	PM

derivative of error Δe (labels for the rows above)

PB : Positive Big   (kf × 2.00)
PM : Positive Medium   (kf × 1.00)
PS : Positive Small   (kf × 0.50)
PV : Positive Very Small   (kf × 0.25)
ZO : Zero   (kf × 0.00)
NV : Negative Very Small   (kf × −0.25)
NM : Negative Medium   (kf × −1.00)

$Kg\,(e,\,\Delta e)_L$

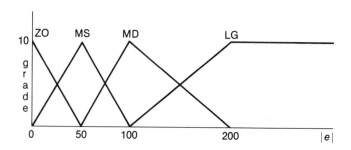

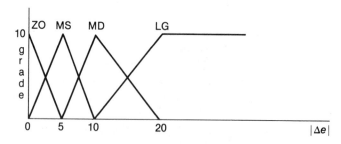

Figure 5.31 Fuzzy rules and membership function

This controller can be easily connected with a large expert computer system that has knowledge-based rules and human beings' experienced knowledge. Then, a computer integrated manufacturing system can be constructed.

BIBLIOGRAPHY

[5.1] Y. Kaku and T. Mita, "Digital Zeroing with Observer," Trans. on IEE Japan, vol. 108-D, no. 7, 1988, in Japanese.

[5.2] S. Z. Sarpturk and O. Kaynak, "Adaptive Pole Placement by Output Feedback Based on the Sliding Mode Control Approach with Application to a D.C. Servo Motor," private communication with the author. His close work is found in Proc. of the IECON'87 (Cambridge, MA, U.S.A.), Nov. 1988.

[5.3] Y. Dote and M. Igarashi, "Digital Signal Processor (DSP)-Based Robust Speed Regulation of D.C. Servo Motor," Proc. of the IECON'87 (Cambridge, MA, U.S.A.), pp. 184–86, Nov. 1987.

[5.4] N. Matsui and H. Ohashi, "DSP-Based Adaptive Control of a Brushless Motor," Conf. Record of the IEEE IAS meeting, pp. 375–81, Pittsburgh, U.S.A., Oct. 1988.

[5.5] A. J. Pollmann, "Software Pulsewidth Modulation for μP Control of A.C. Drives," IEEE Trans. on IA, vol. IA-22, no. 4. July/August 1986.

[5.6] Y. Dote, "Digital Signal Processor (DSP)-Based Variable Structure Control with Equivalent Disturbance Observer for Robot Manipulator," Proc. of the IECON'87 (Cambridge, MA, U.S.A.), Nov. 1987.

[5.7] Y. Dote and M. Shinojima, "DSP-Based Variable Structure PI Controller for Robot Manipulator," in the same Proc. as [5.6], 1987.

[5.8] H. Naito and S. Tadakuma, "Microprocessor-Based Adjustable-Speed D.C. Motor Drives Using Model Reference Adaptive Control," IEEE Trans. on IA, vol. IA-23, no. 2, March/April 1987.

[5.9] K. Ohnishi, et al., "Model Reference Adaptive System Against Rotor Resistance Variation in Induction Motor Drive," IEEE Trans. on IE, vol. IE-33, no. 3, August 1986.

[5.10] M. Nakano, et al., "Sinusoidal Wave Inverter Using Feedback (Repetitive) Control," JICE Trans. vol. 24, no. 7, 1988, In Japanese.

[5.11] Kobayashi, Hara, Tanaka and Nakano, "Motor Speed Fluctuation Reduction Using Repetitive Control," Trans. JEE, D, vol. 107, no. 1, pp. 29–34, 1987.

[5.12] Y. Yabuta, K. Tsujimura, and K. Suzuki, "Force Control by using Fuzzy Set Theory," Proc. of SICE '86 in Japanese; Tokyo, Japan 1987.

[5.13] Y. Dote, "Adaptive Grasping Force Control for Manipulator Hand Using Fuzzy Set Theory," Proc. of the Int. Workshop in Intelligent Robots and System, Tokyo, Oct. 31–Nov. 2, 1988.

Chapter 6

Sensors and Digital Signal Processing

6.0 INTRODUCTION

Controllers and sensors are mutually complementary when they are used as compensators. The sensors plays an important role in the overall control system. This section deals with digital signal processing (including an observer) to obtain a high performance, and low-cost sensing devices. Also, several digital filter processing sensor signals are introduced. A digital signal processor (TMS 320) is suitable for these implementations in both its architecture and programming. The novel speed observers are developed and described by using dual role of a servo motor as a sensor and by using duality between a controller and an estimator.

6.1 HIGH RESOLUTION POSITION-SENSING ALGORITHM*

It is necessary to devise a position-sensing algorithm with high resolution even in the case of obtaining a speed signal with resolution, since the latter is calculated on the basis of the former.

An encoder whose output is a 2-phase sinusoidal wave is used to cope with this. Then the frequency and phase information can be used to get a position signal

---

* See reference [6.1].

with high resolution. The position is given by:

$$\text{Position } \theta = \text{coarse position } \theta_R + \text{fine position } \theta_F$$

The frequency information from the sensor is used to get a coarse position θ_R, and the phase information gives the final position θ_F. This is obtained by measuring the output phase voltages e_A and e_B represented by the following equation:

$$e_A = K \sin \theta_F$$

$$e_B = -K \cos \theta_F$$

e_A and e_B are known. To eliminate K, θ_F is calculated by the DSP.

6.2 HIGH RESOLUTION VELOCITY-SENSING ALGORITHMS

Low-Velocity Observer*

When motor speed is low, the period of the output pulse from the encoder becomes longer than the sampling time in the speed control loop. This results in not obtaining speed information at time of sampling. Having delayed speed feedback signal brings stability problems. To overcome this, it is necessary to estimate the motor speed by using an observer between the encoder pulses. The block diagram of the observer is shown in Fig. 6.1. Every time an encoder pulse comes, the estimate of the load torque is calculated as follows. When the $(i + 1)^{th}$ encoder pulse arrives, the following calculation is carried out by the DSP:

$$\hat{\iota}_L(i + 1) = K_p\{\hat{n}_\ell(i) - n_\ell(i)\} + K_I \sum_{i=1}^{i} \{\hat{n}_\ell(i) - n_\ell(i)\} \qquad (6.1)$$

where

$\hat{\iota}_L(i + 1) \triangleq (i + 1)^{th}$ load torque estimate

$n_1(i) \triangleq$ average motor speed between the ith and $(i + 1)^{th}$ encoder pulses

$\hat{n}_1(i) \triangleq$ estimate of $n_1(i)K_p, K_I \triangleq$ proportional and integral gains,

respectively

Assuming ι_L is constant at the sampling instance, $\hat{n}_f$ is calculated as follows:

$$\hat{n}_f(i, j) = \frac{T_s}{J} \{\iota_r(i, J) - \hat{\iota}_L(i)\} + \hat{n}_f(i, j - 1) \qquad (6.2)$$

$$\hat{n}_\ell(i) = \frac{1}{T_n} \sum_{j=1}^{m} \hat{n}_f(i, j) \qquad (6.3)$$

* See reference [6.2].

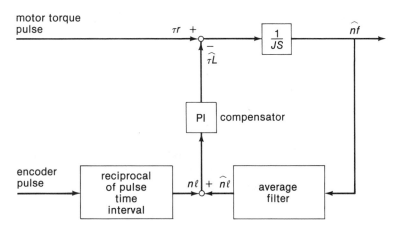

Figure 6.1   Block diagram for low-speed observer

where j represents time elapsed, $T_s \triangleq$ sampling time, after the motor speed is detected, J is an inertia. T_n is encoder pulse interval. Equation (6.2) is used under the assumption that the time constant in a current control loop i is fast enough to neglect in comparison with that in a speed control loop. If a current is measurable, it is more suitable for a motor torque command.

Velocity Observation from Discrete Position Encoder*

If a motor rotates at relatively high speed, the following least squares fit observer is useful.

This method directly computes the derivative of the time as a function of position data which can then be inverted to obtain the velocity at the most recent encoder line, v_K. An approximating polynomial of the form

$$t_K = \sum_{i-0}^{N} C_i K^i \tag{6.4}$$

is assumed, where N is the order of the polynomial fit. When this polynomial is used to describe the Kth and the preceding $M - 1$ samples of t_K, the following system of equations results where the subscripts are dynamically relabeled according to the mapping $K \leftrightarrow M, K - 1 \leftrightarrow M - 1, \ldots, K - M + 1 \leftrightarrow 1$:

$$t = Ac_1 \tag{6.5}$$

where t is an M-vector containing the most recent M time samples, c is an $(N + 1)$-vector containing the polynomial coefficients, and A is an $[Mx(N + 1)]$ −

* See reference [6.3].

matrix with elements $a_{i,j} = i^{j-1}$ where $1 < = i < = M$ and $1 < = j < = N + 1$.

$$t = \begin{bmatrix} t_1 \\ t_2 \\ \vdots \\ t_M \end{bmatrix}, \quad c = \begin{bmatrix} c_0 \\ c_1 \\ \vdots \\ c_N \end{bmatrix}, \quad A = \begin{bmatrix} 1 & 1 & 1 & 1 & & 1^N \\ 1 & 2 & 4 & 8 & \cdots & 2^N \\ 1 & 3 & 9 & 27 & & 3^N \\ & & \vdots & & & \vdots \\ 1 & M & M^2 & M^3 & \cdots & M^N \end{bmatrix} \quad (6.6)$$

When $M > N + 1$, this system of equations is overdetermined. In that case, the coefficient vector c is obtained to minimize the total square error, using standard LSF techniques, as

$$c = (A^T A)^{-1} A^T t = A^+ t \quad (6.7)$$

The derivative dt_K/dx can also be obtained from Eq. (6.4) as

$$\frac{dt_M}{dx} = \sum_{i=1}^{M} c_i(iK^{i-1}) \quad (6.8)$$

Since the coefficients c_1 have already been specified by Eq. (6.7), the derivative of the $t_K th$ sample with respect to index can be written in matrix form as

$$\frac{dt_M}{dx} = q^t A^+ t \quad (6.9)$$

where $q^T = (0\ M\ 2M\ 3M^2 \ldots (N-1)M^{N-2} NM^{N-1})$. This result can be further simplified as

$$\frac{dt_n}{dx} = h^T t \quad (6.10)$$

where $h^T = q^T A^+$. Thus, the derivative dt_K/dx can be obtained as a linear combination of Kth and previous $M - 1$ samples of time as

$$\frac{dt_K}{dx} = \sum_{i=1}^{M} h_i t_{K-M+i} \quad (6.11)$$

where the $h_i s$ are the M coefficients of the h^T vector and have been calculated off-line to implement the LSF observer. The observer output is the reciprocal of the velocity, the derivative of the least squares fit polynomial of $t(x)$. This observer can be implemented as a finite impulse response digital filter of order M.

This is implemented with a DSP (TMS 320C25) in the author's laboratory. The first order polynomial is the most suitable for this curve fitting. This has been confirmed by experiments.

6.3 DISTURBANCE TORQUE OBSERVER

Equivalent Disturbance Observer

In section 5.1, the following observer is derived. The motor torque equation can be written as:

$$sJw(s) + T_\ell(s) = K_t i_a(s) \qquad (6.12)$$

where $J \triangleq$ inertia
$w \triangleq$ motor speed
$T_\ell \triangleq$ external load torque, including friction force
$K_t \triangleq$ torque constant
$i_a \triangleq$ armature current

Let $J = \hat{J} + \Delta J$ and $K_t = \hat{K}_t + \Delta K_t$, $(\hat{\ })$ represents the nominal value and Δ shows the deviation from the nominal value.

Let the equivalent distribution be T_e. Then:

$$T_e(s) = T_\ell(s) + s\Delta Jw(s) - \Delta K_t i_a(s) \qquad (6.13)$$

From Eqs. (6.12) and (6.13)

$$s\hat{J}w(s) + T_e(s) = \hat{K}_t i_a(s)$$

or

$$T_e(s) = s\hat{J}w(s) + \hat{K}_t i_a(s)$$

To obtain the estimate of $T_e(s)$, $\hat{T}_e(s)$, the following observer is constructed by using a low pass filter $[1/(\hat{J}_s + K)]$:

$$\hat{T}_e(s) = [-s\hat{J}w(s) + \hat{K}_t i_a(s)]/(\hat{J}S + K) \qquad (6.14)$$

The block diagram for this observer is shown in Fig. 6.2. The observer is used as a compensator for equivalent disturbance cancellation. The observer

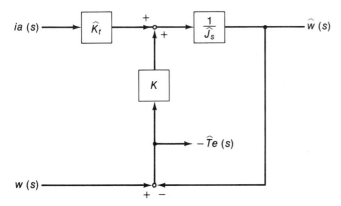

Figure 6.2 Block diagram for equivalent disturbance observer

structure is similar to the one in Fig. 6.1. Equation (6.14) is implemented with a DSP by using the bilinear transformation and the frequency prewarping described in Chapter 7.

Load Torque (Constant) Observer with Digital Filters

A constant load torque observer is derived as follows:

$$Jw(s) = -T_l(s) + K_t i_a(s) + K_1[_w(s) - w(s)] \tag{6.15}$$
$$S_{T\ell}(s) = K_2[_w(s) - w(s)]$$

where $J \triangleq$ inertia
$T_\ell \triangleq$ constant load torque
$K_t \triangleq$ torque constant
$i_a \triangleq$ armature current
$K_1, K_2 \triangleq$ observer gains
$\hat{w}, \hat{T}_e \triangleq$ estimate of w and T_e

The block diagram for this observer is illustrated in Fig. 6.3. This is close to the one in Fig. 6.1. This observer is used as a compensator for motor speed drop due to sudden load torque variation. It is implemented with a DSP by using the transformations described in Chapter 7.

Then, moving average digital and variable structure filters are also included in the load torque observer, as shown in Fig. 6.4. The moving average filter eliminates the torque pulsation component caused by converter arm firing angle

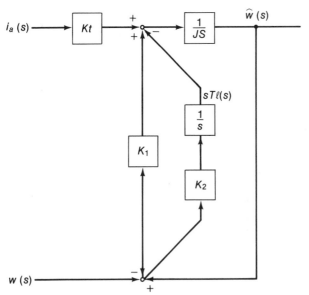

Figure 6.3 Block diagram for constant torque observer

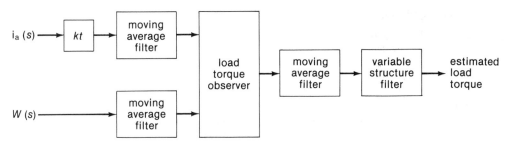

Figure 6.4 Digital filters used for load torque estimation

inequalities, and the variable structure filter reduces noise in the estimated load torque.

Moving Average Digital Filter

The noise component contained in the input signal of the observer consists of the harmonic components of AC source frequency, due to inequality in the firing angles of the converter arms and noise in the speed and current detecting circuit. When these noises enter the estimated load torque, it impairs the stability of the system. The moving average filter is used to eliminate noise caused by the harmonic components. The moving average digital filter is expressed as follows:

$$y(i) = \frac{1}{\sum_j w(j)} \sum_{j=1}^{N} u(i - j)w(j) \tag{6.16}$$

where y = output
   N = number of points taken, averaged at the ith instant
   u = input
   w = weighting sequence

This is easily implemented with a DSP.

Variable Structure Digital Filter

Noise in the detecting circuit contains low- to high-frequency components, so it cannot be eliminated by the moving average filter alone; hence, a first-order filter with variable structure (variable filter) is employed.

The state equation of the variable filter is:

$$q(i + 1) = q(i) + K_f(i)e(i) \tag{6.17}$$
$$e(i) = p(i) - q(i)$$

where $q \triangleq$ output
$\quad p \triangleq$ input
$\quad k_f \triangleq$ variable filter gain
$\quad e \triangleq$ error

The block diagram for this filter is shown in Fig. 6.5.

The value of K_f can be adjusted as follows: When the load torque changes from T_{1a} to T_{1b} in step as shown in Fig. 6.6(a), the value of T_1 contains noise mixed with the observer input signal and noise caused by harmonics, in addition to the estimated signal component [see Fig. 6.6(a)]. When the signal component of T_1 is T_{1a} in the steady state and diverges to T_{1b} after transient, the value of T_1 fluctuates in the vicinity of the steady-state value, due to the noise distributed in the range of $-e_1$ to e_1. So, the value of K_f is minimized by the variable filter to eliminate the noise when the signal component is diverged, while it is increased only at the change of T_1 to feed the signal. In this case, the component contained in the filter input/output difference, e at the transient and steady state is divided into signal and noise. Using a fuzzy method, K_f is obtained from their probability distributions as shown in Fig. 6.6(b).

In this way, a pattern of K_f, which varies with e, can be obtained as shown by the dotted line in Fig. 6.6(c). In order to make it simple, a segmented line pattern, which varies with $+/-e_1$ and $+/-e_2$ is used. In area I [Fig. 6.6(c)], this pattern is based on $K_{f\text{min}} = g_2/g_1$ so that the filter's pole cancels with the zero of the observer, and in area II, the value of K_f is varied in straight line, while in area III, $K_f = K_{f\text{max}} < = 1$ (no delay of filter at 1 to prevent hunting).

In this case, the value of e_2 is set to $e_2 = > 2e_1$ so that area II's width becomes larger than area I's, when the noise or signal in area I varies from $-e_1$ to e_1. With such a pattern, the gain varies continuously with changes in the signal level to provide stable operation.

In general, there exists some delay time when applying digital filtering to the observer (an observer is itself a digital filter). The following digital lead time

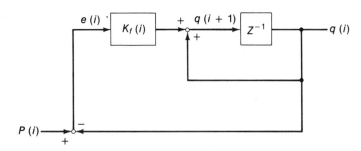

Figure 6.5 Block diagram for variable structure digital filter

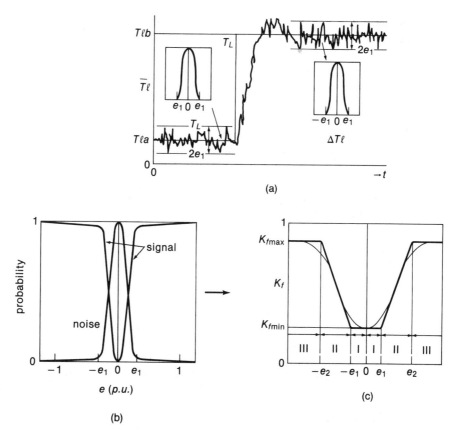

Figure 6.6   Tuning method of gain K_f. (a) Step response of Tl containing noise.
(b) Components of e. (c) Gain K_f pattern.

filter is used to cancel the effect of the delay time:

$$y(i) = u(i) + \lambda[u(i + 1) - u(i)] \tag{6.18}$$

where $y \triangleq$ output
     $u \triangleq$ input
     $\lambda \triangleq$ filter gain

Digital filters are implemented with DSPs.

6.4 DISCUSSIONS ON OBSERVERS

The observers derived in this chapter are designed by using dual role of a servo
motor as a sensotor.

    There exists duality between control (controller) and estimation (observer).

In this section, the robust algorithms described in Chapter 5 are applied to speed observer designs by using a position sensor (encoder).

Robust speed observer 1.   First, the zeroing with an equivalent observer in section 5.1.1 is used. Equation (5.47) can be rewritten as

$$L\dot{i}_a + ri_a + k_\omega\dot{\theta} = kV_{in} + d \qquad (6.19)$$

In the same way, an equivalent disturbance observer is designed as

$$\hat{T}_e(s) = (\hat{k}V_{in}(s) - (s\hat{L} + \hat{R})i_a(s) - s\hat{k}_\omega\theta(s)/\hat{L}_s + k + \hat{R})$$

The block diagram for this observer is shown in Fig. 6.7.

$T_e(s)$ is feedforward into the input $V(s)$ to cancel Eq. (6.20), the equivalent disturbance. Then, the following equation is obtained from Eq. (6.19);

$$(s\hat{L} + \hat{R})i_a(s) + \hat{k}_\omega w(s) = \hat{k}V_{in} \qquad (6.20)$$

Now

$$w(s) = \frac{1}{\hat{k}_\omega}[\hat{k}V_{in} - (s\hat{L} + \hat{R})i_a(s)] \qquad (6.21)$$

Equation (6.21) can be easily calculated from the observer information (* and ** in Fig. 6.7). Thus, robust speed estimation is obtained.

Caution: this method is provided only for explanation. It is not implemented. In [6.5], an adaptive observer is described; [6.5] is recommended.

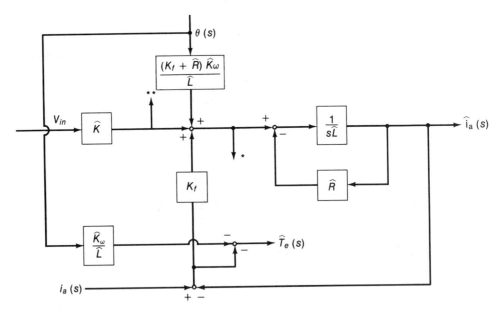

Figure 6.7 Equivalent disturbance observer

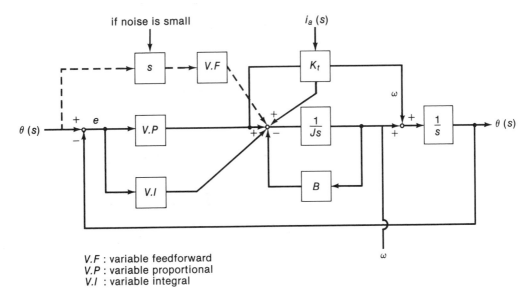

V.F : variable feedforward
V.P : variable proportional
V.I : variable integral

Figure 6.8   Variable structure PI observer

Robust speed observer 2.   Next, the variable structure PI algorithm described in section 5.3 is used for fast estimation error convergence.

The observer described in Fig. 6.2 is modified. The block diagram is shown in Fig. 6.8 and Fig. 5.11. This is a nonlinear observer. This variable PI algorithm is also applied to the filter described in Fig. 6.5, and it can separate the noisy signal into the small and the large signal. A linear observer is found in [6.6].

BIBLIOGRAPHY

[6.1] Y. Sugiura et al., "Digital Speed Detection Method with High Resolution for Servo Motor," Trans. of the JEE, vol. 108, no. 1, ptd. 1988 in Japanese.

[6.2] M. Watanabe et al., "Digital Servo System with Speed Estimation Observer," Trans. of the JEE, vol. 107, no. 12, ptd. 1987 in Japanese.

[6.3] R. H. Brown and S. C. Schneider, "Velocity Observations from Discrete Position Encoders," Proc. of the IECON '87, Cambridge U.S.A., Nov. 1987.

[6.4] H. Umida and M. Ohara, "A Digital Observer for Impact Drop Compensation," Proc. of the IECON '87, Cambridge U.S.A., Nov. 1987.

[6.5] J. P. Rognon, D. Roye, and D. S. Zhu, "A Simple Speed Observer for Digitally Controlled Motor Drives at Low Speed," IEEE IAs Conf. Record, pp. 369–74, Pittsburgh U.S.A., Oct. 1988.

[6.6] R. D. Lorentz and K. V. Patten, "High Resolution Velocity Estimation for All Digital A.C. Servo Drivers," in the same Conf. Record as [6.5], pp. 363–68.

Section III
System Implementation

Chapter 7

Implementation Issues of Digital Servo Motor and Motion Control Algorithms with a TMS 320

7.0 INTRODUCTION

Most of this chapter is based on Dr. Slivinsky's work (see reference [7.1]). Implementation issues are described by using TMS 32010; however, they are applicable to TMS 320C25.

7.1 CONSIDERATION OF ERRORS IN DIGITAL CONTROL SYSTEMS

There are practical limitations in microprocessor applications. Finite work length (resolution or quantization error), and time delay due to computation and sampling time encountered in the execution of the operational instructions of the processor, must be taken into account by the control designer. They have important effects on the performance of the control system. The high sampling rate results in low sensor signal resolution. This increases quantization error and brings in system stability problems; however, the low sampling rate causes delay time, giving poor system stability, though it yields precise information and enough signal processing time. It is recommended to use simulation techniques in order to determine their optimal values, since they are mutually affected and complicated. However, in the usage of DSPs, digital compensator algorithms execute processors (TMS 320 family) that use finite precision arithmetic. The signal-quantization errors associated with finite-precision computations and the methods for the handling of these errors are important, and are presented next.

Fixed-Point Arithmetic and Scaling

Computation with the TMS 32010 is based on the fixed-point two's-complement representation of numbers. Each 16-bit number has a sign bit, i integer bits, and 15-i fractional bits. For example, the decimal fraction $+0.5$ may be represented in binary as

$$0.100\ 0000\ 0000\ 0000$$

This is Q15 format since it has 15 fractional bits, 1 sign bit, and no integer bits. The decimal fraction $+0.5$ may also be represented in Q12 format as

$$0000.1000\ 0000\ 0000$$

This number is in the Q12 format because it has 12 fractional bits, 1 sign bit, and 3 integer bits. Note that the Q15 notation allows higher precision, while the Q12 notation allows direct representation of larger numbers.

For implementing signal-processing algorithms, the Q15 representation is advantageous, because the basic operation is multiply-accumulate and the product of two fractions remains a fraction with no possible overflows during multiplication. When the Q12 format is used, a software check for overflow is necessary. Section 7.13, Overflow and Underflow Handling, provides a detailed analysis of overflow handling.

In the case where two numbers in Q15 are multiplied, the resulting product has 30 fractional bits, 2 sign bits, and (as expected) no integer bits. To store this product as a 16-bit result in Q15, the product must be shifted left by 1 bit and the most-significant 16 bits stored. The TMS32010 instruction SACH allows for this 1-bit shift.

In the case where a Q15 number is to be multiplied by a 13-bit fractional signed constant represented as a Q12 number, the result (to correspond with Q15) must be left-shifted 4 bits to maintain full precision. The TMS32010 instruction SACH allows for the appropriate shift.

When fixed-point representations are used, the control system designer must determine the largest magnitudes that can occur, for all variables involved in the computations required by the digital compensator. (Floating-point representations allow larger magnitudes, but take more time for the microprocessor to perform the required computations.) Once these largest magnitudes are known, scaling constants can be used to attenuate the compensator input as much as necessary, to ensure that all variables stay within the range that can be expressed in the given representation.

Several methods are used to determine bounds on the magnitudes of the variables. One method, called upper-bound scaling, provides a useful bound on the magnitude. It is sometimes too conservative but it is straightforward to calculate. Consider a variable $y(n)$ that is obtained as the output of a digital compensator $H(z)$ when the input is the sequence $x(n)$. The bound on $y(n)$ is given by

$$y_{\max} = | x_{\max} | \sum_{n=1}^{\infty} | h(n) | \qquad (7.1)$$

where $x_{\max}$ is the maximum value in $x(n)$ and the sequence $h(n)$ is the unit-sample response sequence of the digital compensator $H(z)$. Other methods for estimating the upper bound are L_p-norm scaling, unit-step scaling, and the averaging method.

After $y_{\max}$ is determined, the scale factor can be chosen as the multiplier that is applied to $x(n)$ prior to the compensator computations to ensure that $y(n)$ remains within the required bounds. In addition, the control system designer may have knowledge concerning the bounds on the compensator variables based on prior experience, the characteristics of the corresponding variables of analog prototypes, and simulation results.

Finite-wordlength Effects

All variables involved in the digital compensator—the input, the compensator coefficients, the intermediate variables, and the output—are represented as finite-wordlength numbers. This restriction gives rise to errors. Another source of errors is the truncation or rounding that takes place when the 32-bit product of two 16-bit numbers is stored as a 16-bit number. Both of these errors give rise to the finite-wordlength effects discussed in this section.

The representation of the compensator input as a finite-precision (quantized) number produces an input-quantization error. The size of this error, for a rounding scheme, can be anywhere from $-(2^{-B})/2$ to $(2^{-B})/2$, where B is the number of bits in a word. The input-quantization error is usefully modeled as a zero-mean random variable, uniformly distributed between its positive and negative bounds. A technique is available to calculate the variance of the corresponding error at the compensator output (its mean is zero). In this manner, the designer can determine the effect of input quantization on the compensator output.

Similar quantization errors are associated with the multiplication process. Each multiplication is assumed to produce the "true" product with an error that is a zero-mean, uniformly distributed random variable. The variance of the corresponding error at the compensator output can be calculated in the same manner as for the error, due to input quantization. These individual variances are then added to measure the total effect at the compensator output for each truncation or rounding.

Another way to describe the effects of truncation or rounding is in terms of "limit cycles," which are sustained oscillations in the closed-loop system. These oscillations are caused by nonlinearities within the loop. In this case, the nonlinear quantizations are associated with the multiplications. Limit cycles persist even when the system input goes to zero, and their amplitude can be sizable. No general theory is available to treat this nonlinear phenomenon. Bit-level simulations which model the compensator and the complete closed-loop system are used to ascertain their presence and effect on the closed-loop performance.

When a digital compensator is implemented as an algorithm, to be executed on finite-precision hardware, a problem arises with implementing the coefficients present in the corresponding transfer function (see section 7.4, TMS 320 Implementation of Compensators and Filters). The infinite-precision compensator coefficients must be rounded and stored, using a finite-length, fixed-point binary representation. Due to this coefficient-quantization effect, the performance of the implemented filter will deviate from the design performance.

The deviation in performance can be estimated by computing the filter's pole and zero locations and the corresponding frequency response magnitude and phase for the compensator with the quantized coefficients. Coefficient quantization forces the filter's pole and zeros into a finite number of possible locations in the z-plane and is of most concern for filters with stringent specifications, such as narrow transition regions.

The designer must choose the filter structure least sensitive to inaccurate coefficient representation. The choice should be of a modular rather than a direct filter structure. For example, a higher order filter should be implemented as a cascade or parallel combination of first-order and second-order blocks. The reason for this choice is the lesser sensitivity to coefficient variations of the roots of low-degree polynomials in comparison with high-degree polynomials. Several methods for selecting the filter structures least affected by coefficient quantization are available.

To quantitatively evaluate the effect of coefficient quantization of the position of the poles or zeros of a digital transfer function, a "root sensitivity function" can be computed.

Overflow and Underflow Handling

Digital control system algorithms are usually implemented using two's-complement, fixed-point arithmetic. This convention designates a certain number of integer and fractional bits. The fixed-point arithmetic computations may, at some point, produce a result that is too large to be represented in a chosen form of fixed-point notation (e.g., Q12). The resulting overflow, if untreated, may cause degraded performance, such as limit cycles and large noise spikes at the filter's output, which may contribute to the system's instability. The system must be able to recover from the overflow condition, i.e., return to its normal nonoverflow state.

Consider an example of the Q12 representation: The number 7.5 multiplied by itself gives the result of 56.25; however, in an overflow in Q12 no hardware overflow occurs in the accumulator; i.e.,

$$
\begin{array}{rl}
7.5 & 0111.1000\ 0000\ 0000 \\
\times\ 7.5 & 0111.1000\ 0000\ 0000 \\
\hline
56.25 & 0011\ 1000.0100\ 0000\ 0000\ 0000\ 0000\ 0000
\end{array}
$$

For the Q12 representation, the above 32-bit product is shifted left 4 bits and the leftmost 16 bits are retained:

$$1000.0100\ 0000\ 0000$$

The correct answer is 56.25, but the number stored in the Q12 representation is -7.75.

The TMS 32010 has a built-in overflow mode of operation that, if enabled, causes the accumulator to saturate upon detection of an overflow, during addition when the accumulator register overflows. During multiplication, an overflow of the fixed-point notation may also occur even though the hardware overflow of the accumulator register does not occur. This is because the 32-bit result of a multiplication of two 16-bit numbers must be stored in a 16-bit memory word, in the form consistent with the chosen fixed-point notation (see above example). To adjust the location of the binary point, the storing operation requires that the number in the accumulator be shifted left and truncated on the right before storing. If the most significant bits shifted out contain magnitude information, in addition to sign information, an overflow in the chosen fixed-point notation results.

To track overflows associated with the number representation, the control system software should contain an appropriate overflow-checking routine in those places where multiplications and additions occur. This routine should not rely exclusively on the TMS 32010's overflow mode to intercept and correct the overflow occurrences.

Two approaches may be used to handle overflows. The first is to prevent the overflow from occurring by choosing conservative scaling factors for the numbers used in computations, as described in section 7.1.1, Fixed-Point Arithmetic and Scaling. These scaling factors are used to limit the range of inputs to each of the basic building blocks of the compensator, namely, the first- and second-order filter sections. The scaling factor chosen reduces the input magnitude and, consequently, all other signal levels, thereby enabling the compensator coefficients, the expected inputs, and their products and sums, to all be represented without overflow. The scaling must also maintain the signal levels well above the quantization noise.

The second approach for handling overflow is to adjust the sum or product each time an overflow occurs. To accomplish this, an overflow checking routine must be written and executed at certain points along the computational path. The routine must check whether the number just computed and residing in the 32-bit accumulator can be stored without overflow in a 16-bit memory location in accord with the chosen fixed-point notation. Once the routine detects an overflow condition, it should replace the computed number with the maximum or minimum representable two's-complement number. This scheme simulates a saturation condition present in analog control systems. To prevent overflow limit cycles, the saturation overflow characteristic is preferred to the two's-complement, "wrap-around" characteristic.

An example of the overflow checking and correcting technique, for a first- and second-order filter subroutine, is provided in section 7.4, TMS 320 Implementation of Compensators and Filters. This Direct-Form II implementation subroutine checks for overflow occurrences upon computation of the filter's intermediate state variable and again upon computation of the filter's output.

In a digital control system, the first- and second-order building blocks are either cascaded or connected in parallel to compute a series of control algorithms. The first- and second-order filter subroutines, called to compute each of the control system's elements, use 16-bit memory locations as storage media for their intermediate values, in which case, it is appropriate to check for overflow in each block.

At the end of a computational chain—before the final computed digital output is ready for transfer to the analog domain—it is necessary to check that the number being sent to the digital-to-analog converter is within the range, based on the manner in which the converter is interfaced to the processor data bus. For example, if a 12-bit converter is wired to the 12 least significant bits (LSBs) of the 16-bit processor data bus, then the 12 LSBs must contain both magnitude and sign information, which may require that the original 16-bit number be adjusted or limited before being sent to the converter.

Underflow conditions, which can also appear during digital control algorithms, are conceptually similar to overflows in that the computed value contained in the 32-bit accumulator is too small to be accurately represented in a 16-bit memory word in the chosen fixed-point notation. One possible solution to this problem is to multiply the small result by a gain constant, to raise its value to a representable level. The appropriately chosen gain constant may come as a result of gain distribution throughout the digital control system, whereby large gains from some of the building blocks get uniformly distributed over a range of the system's sections.

7.2 DISCRETIZATION OF CONTINUOUS SYSTEMS (CONTROLLER DESIGN)

Slow Dynamic Systems (Design Based on Analog Prototype)

A commonly used method of designing a digital control system is to first design an equivalent analog control system, using one of the well-known design procedures. The resulting analog controller (analog prototype) is then transformed to a digital controller by the use of one of the transformations described below.

The design of the analog controller may be carried out in the s-plane using design methods such as root-locus techniques, Bode plots, the Routh-Hurwitz criterion, state-variable techniques, and other graphic or algebraic methods. The purpose is to devise a suitable analog compensator transfer function, which is transformed to a digital transfer function. This digital transfer function is then inverse z-transformed to produce a difference equation that can be implemented as an algorithm, to be executed on a digital computer. Two of the analog-to-digital transformation methods, the matched pole-zero and the bilinear transformation, are described as follows:

1. The matched pole-zero (matched Z-transform) method, maps all poles and zeros of the compensator transfer function from the s-plane to the z-plane according to the relation:

$$z = e^{sT} \tag{7.2}$$

where T is the sampling period. If more poles than zeros exist, additional zeros are added at $z = -1$, and the gain of the digital filter is adjusted to match the gain of the analog filter at some critical frequency (e.g., at DC for a low pass filter). This method is somewhat heuristic and may or may not produce a suitable compensator.

2. The bilinear (Trustin) transformation method approximates the s-domain transfer function with a z-domain transfer function by use of the substitution:

$$s = \frac{2}{T} \frac{z-1}{z+1} \tag{7.3}$$

As in the matched pole-zero method, the bilinear transformation method requires substitution for s. Compensators in parallel or in cascade maintain their respective structures when transformed to their digital counterparts. This substitution maps low analog frequencies into approximately the same digital frequencies, but produces a highly nonlinear mapping for the high frequencies. To correct this distortion, a frequency prewarping scheme is used before the bilinear transformation. The frequency prewarping operation results in matching the single critical frequency between the analog domain and the digital domain. To achieve this result, the prewarping operation replaces each s in the analog transfer function with $(\omega_o/\omega_p)s$, where ω_o is the frequency to be matched in the digital transfer function and

$$\omega_p = \frac{2}{T} \tan \frac{\omega_o T}{2} \tag{7.4}$$

Bilinear transformation with frequency prewarping provides a close approximation to the analog compensator and is the most commonly used technique. Other methods for converting a transfer function from the analog to the digital domain are: the method of mapping differentials, the impulse-invariance method, the step-invariance method, and the zero-order hold technique.

The basic disadvantage of design, based on an analog prototype, is that the discrete compensator is only an approximation to the analog prototype. This analog prototype is an upper bound on the effectiveness of the closed-loop response of the digital compensator. The numerical example is given next.

Development of a Digital Compensator Transfer Function

The development of a digital equivalent of an analog compensator transfer function using the bilinear transformation with frequency prewarping is shown in this section.

Beginning with an analog prototype transfer function:

$$G(s) = 1000 \frac{S^2 + 68.2S + 3943}{S^2 + 2512S + 6.31 \times 10^6}$$

The sampling frequency to be used in converting to a digital equivalent is $f = 4020$ Hz (i.e., the sampling period $T_s = 1/4020s = 248.76 \times 10^{-6}s$).

The characteristic equation of this analog transfer function is

$$S^2 + 2512S + 6.31 \times 10^6 = 0$$

which fits the standard, second-order form:

$$S^2 + 2\zeta\omega_n S + \omega_n^2 = 0$$

The natural frequency $\omega_n = \sqrt{6.31 \times 10^6} = 2511.9713$ rad/s. To compensate for nonlinear mapping of analog-to-digital frequencies by the bilinear transformation method, the natural frequency is prewarped according to the formula:

$$\omega_p = \frac{2}{T} \tan \frac{\omega_o T}{2}$$

$$= \frac{2}{248.76 \times 10^{-6}} \tan \frac{2511.9713 \times 248.76 \times 10^{-6}}{2}$$

$$= 2597.03 \text{ rad}/s$$

This prewarping scheme matches exactly the natural frequency in the analog and digital domains for the compensator.

To obtain the prewarped version of the analog transfer function, s is replaced with $(\omega_o/\omega_p)s$. It is therefore convenient to compute the ratio:

$$\frac{\omega_o}{\omega_p} = \frac{2511.9713}{2597.03} = 0.9672$$

The prewarped $G(s)$, i.e., $G_p(s)$, is then computed as

$$G_p(s) = 1000 \frac{(0.9672s)^2 + 68.2(0.9672s) + 3943}{(0.9672s)^2 + 2512(0.9672s) + 6.31 \times 10^6}$$

$$= 1000 \frac{s^2 + 70.51s + 4214.87}{s^2 + 2597.16s + 6.75 \times 10^6}$$

Bilinear transformation is next applied to $G_p(s)$ whereby the continuous variable s is replaced by the expression that involves the discrete variable z:

$$S = \frac{2}{T} \frac{z - 1}{z + 1}$$

This produces the discrete transfer function $D(z)$. For the compensator:

$$D(z) = G_p(s) \Big|_{s = \frac{2}{T}\frac{z-1}{z+1}} = \left(1000 \frac{s^2 + 70.51s + 4214.87}{s^2 + 2597.16s + 6.75 \times 10^6} \right)\Big|_s = \frac{2}{248.76 \times 10^{-6}} \frac{z - 1}{z + 1} \right)$$

After further computations:

$$D(z) = 706.76 \frac{1.0 - 1.9824z^{-1} + 0.9826z^{-2}}{1.0 - 1.2548z^{-1} + 0.5474z^{-2}}$$

The final step is the gain adjustment in the digital transfer function. This can be accomplished by matching the analog and digital gains at some predetermined frequency, for example, DC.

For the DC case, $s = j\omega = 0$, and from the bilinear transformation:

$$z = \frac{2 + sT}{2 - sT} = 1$$

Therefore, at DC, $G(0) = D(1)$.

For this transfer function, $G(0) = 0.6249$, $D(1) = 0.5072$. If $G(0) = KXD(1)$, then the constant K becomes

$$\frac{0.6249}{0.5072} = 1.2321$$

The final form of the digital equivalent transfer function is

$$D(z) = 870.77 \frac{1.0 - 1.9824z^{-1} + 0.9826z^{-2}}{1.0 - 1.2548z^{-1} + 0.5474z^{-2}}$$

where the gain of 870.77 is the product of Kx (the unadjusted digital gain), i.e., $870.77 = 1.2321 \times 706.76$.

Fast Dynamic Systems

The design of a controller for a slow processor (for a fast dynamic system) is as follows:

For delay time compensation, a predictive control with an observer is used. Let's consider the following continuous system (a motor drive system is represented by the state equation).

$$\dot{x} = A_c x + B_c u \tag{7.5}$$
$$y = C_c x$$

With a 0-order holder, it is discretized as follows:

$$x(i + 1) = Ax(i) + Bu(i)$$

$$y(i) = x(i) \tag{7.6}$$

where

$$A = \exp[A_c T], \ B = \int_0^T \exp[A_c T] \, dt \, B_c$$

The following performance index J is assigned.

$$J = \sum_{i=0}^{\infty} \{x(i)^T Q x(i) + u(i)^T R u(i)\} \tag{7.7}$$

We obtain $u(i)$ such that J is minimized by solving the Riccati equation:

$$u(i) = Fx(i) \tag{7.8}$$

This $u(i)$ is not implementable since observation and control should be given at the same time; in other words, there is no computational time to generate the input control.

Where

$$F = (R + B^T PB)^{-1} B^T PA \tag{7.9}$$
$$P = Q + A^T PA - A^T PB(R + B^T PB)^{-1} B^T PA$$

Then, a closed-loop control system is

$$x(i + 1) = (A - BF)x(i) \tag{7.10}$$

Since

$$x(i) = Ax(i - 1) + Bu(i - 1) \tag{7.11}$$

$u(i)$ can be rewritten as follows:

$$u(i) = -FAx(i - 1) - FBu(i - 1)$$

$$u(i + 1) = -FAx(i) - FBu(i) \tag{7.12}$$

$$\begin{bmatrix} x(i + 1) \\ u(i + 1) \end{bmatrix} = \begin{bmatrix} A & B \\ -FA & -FB \end{bmatrix} \begin{bmatrix} x(i) \\ u(i) \end{bmatrix} \tag{7.13}$$

Now, one sampling period can be saved for computation time.

7.3 SOFTWARE CONSIDERATIONS IN IMPLEMENTING DIGITAL SERVO MOTOR CONTROL WITH A TMS 320

To maximize the manageability and portability of the system software, a modular or top-down design technique should be used. This section shows how the modular software structure and the proper layout of system memory contribute to the efficient implementation of a digital control design.

Modular Software System

The concept of modular software design is a technique developed to make system software more manageable and portable. Top-down design is used to break up a large task into a series of smaller tasks or building blocks, which in turn are used for structuring a total system in a level-by-level form. At the end of a top-down design process, a number of modules are linked together which, under the control of the main program, perform as a complete system.

In addition to making the software-development and software-modification processes more manageable, modular design also enhances software portability. Digital control systems use a number of standard functional blocks, such as compensators, notch filters, and demodulators. It is therefore likely that a designer who already has access to one digital control system will want to borrow some of its functional building blocks to quickly implement a new, different control unit, or reconfigure the existing one. The designer who has access to these functional blocks or modules needs only to modify the main program by providing a different sequence of subroutine calls. An initialization routine, a first- and second-order filter routine, a round-off routine, and an overflow checking routine, are examples of functional building blocks.

Each software module is written as a subroutine with a clear and efficient interface (for parameter passing, stack use, etc.) with the main program. In order to maintain the general-purpose function of the module, the data used in computations within a module (i.e., filter coefficients, state-variable values, etc.) should be accessed using indirect addressing, rather than direct addressing. Only those variables whose values remain unchanged should be addressed directly.

Layout of TMS 32010 Data Memory

The layout of the TMS 32010 data memory in a digital control system implementation should be defined in accordance with the requirements of the software modules used in the implementation of the system. The procedure is illustrated by the first- and second-order filter subroutine of section 7.4, TMS 32010 Implementation of Compensators and Filters. This subroutine manipulates its pointer registers so that upon completion of the computations in one filter section, the registers automatically point to the set of coefficients and state variables of the next filter coefficient and state-variable sections, in the order of execution of the control-system algorithms; a sequence of compensators and filters may be executed with a single subroutine call for each element. This scheme enables faster execution of the control algorithms since there is no need to explicitly reload the pointers in order to match the requirements of the current software module being called.

A designer must define all of the data memory locations (set up a system memory map) at the beginning of the program. An efficient way to accomplish this is to use the TMS 32010 assembler's DORG (dummy origin) directive. This directive does not cause code generation. DORG defines a data structure to be used by the system; i.e., it generates values corresponding to the labels of consecutive data memory locations. Using the DORG directive, as opposed to equating labels with data memory locations through the EQU directive, provides flexibility when the data structure needs to be modified. For example, when defining a number of new data memory locations, the labels are inserted in the middle of the "dummy" block and the assembler assigns the values automatically. This function would have to be performed manually if the EQU directive were used.

The software designer must also build a table in the TMS 32010 program memory that corresponds to the previously defined data memory map. The table is then loaded into the data memory by the initialization routine during system startup.

These techniques are illustrated in section 7.4, TMS 32010 Implementation of Compensators and Filters. Note that in the example program in Appendix B, location ONE has to be the last location in the table. Note also that the states and the coefficients of the filters are defined in reverse order; to the order in which the filter executes. This is due to the way the initialization and filter routines are written.

7.4 TMS 320 IMPLEMENTATION OF COMPENSATORS (CONTROLLERS) AND FILTERS

Design procedure and error handling for the standard first- and second-order compensator and filter subroutine are described in this section. Methods for implementing higher order structures, implementation trade-offs, and examples of typical compensators and filters are also given.

Standard First-Order and Second-Order Block as a Subroutine

A standard first- and second-order compensator section is a prime example of the building block philosophy discussed earlier. The routine presented here computes first- and second-order IIR filter sections, using the Direct-Form II network structure, and performs roundoffs and overflow checking. The Direct-Form II, although it somewhat obscures the definition of the variables, is chosen over the Direct-Form I because it requires fewer "delays," i.e., data storage locations, in its computational algorithm.

Consider the second-order transfer function:

$$D(z) = \frac{NO + N1z^{-1} + N2z^{-2}}{1 + D1z^{-1} + D2z^{-2}} = \frac{U(z)}{E(Z)} \tag{7.14}$$

For Direct-Form II, the corresponding difference equations are

$$x(n = c(n) - D1x(n - 1) - D2x(n - 2) \tag{7.15}$$
$$u(n) = NOx(n) + N1x(n - 1) + N2x(n - 2)$$

The signal flowgraph for this transfer function is shown in Fig. 7.1.

The filter routine accommodates a scaling scheme as defined on the main program level; i.e., the values can be scaled by 2^{15} or 2^{12}. The routine is written so that chain implementation of a number of compensators and filters is possible, if the data structure, namely, the coefficient and state-variable tables, are properly arranged. The routine takes its input from the accumulator and outputs the result to the accumulator, so that the main program can efficiently call for the successive execution of the filter routine with a different set of parameters each time.

Two checks for overflow are made within the filter routine. One is made upon computing the value of the intermediate (state) variable, and the other, upon computing the filter's output. Each overflow check determines whether the 32-bit computed result can be stored in a 16-bit memory location, under the adopted scaling scheme. If an overflow condition occurs, the routine "saturates" the output; i.e., it returns the maximum or minimum representable value.

A drawback of overflow checking upon computing the output is the loss of precision of the least-significant bits of the accumulator, which are truncated during accumulator storing operation. This loss of precision is insignificant, however, in comparison with the loss of precision due to an overflow condition.

The first-order filter section is computed exactly like the second-order section. The two coefficients $N2$ and $-D2$ that multiply the "oldest" value of the intermediate (state) variable, i.e., they are equal to zero. This scheme reduces the second-order digital filter to the first-order filter. An example program that uses the first- and second-order filter routine to compute several elements of a digital control system is given later.

Higher Order Filters: Trade-offs Between Cascade and Parallel Types

A higher order filter or compensator in a digital control system can be implemented either as a single section or as a combination of first- and second-order sections. The single section or direct implementation form is easier to implement and executes faster, but it generates a larger numerical error. The larger error occurs because the long filter computation process involves a substantial accumulation

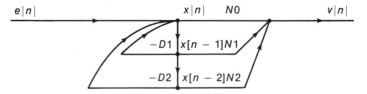

Figure 7.1 Direct-form II compensator/filter

of errors, resulting from multiplications by quantized coefficients and because the roots of high-order polynomials are increasingly sensitive to changes in their (quantized) coefficients. For these reasons, the direct realization form is not recommended except for very low-order controller.

The suggested method of implementing a high-order transfer function is to decompose it into first-order blocks (to accommodate complex conjugate poles or pairs of real poles), and connect these blocks either in a cascade or a parallel configuration.

For the cascade realization (see Fig. 7.2), the transfer function must be decomposed into a product of first-order and second-order functions of the form:

$$D(z) = KD_1(z)D_2(z) \ldots D_n(z) \tag{7.16}$$

Each second-order block has the form:

$$D(z) = \frac{NO + N1z^{-1} + N2z^{-2}}{1 + D1z^{-1} + D2z^{-2}} \tag{7.17}$$

Each first-order block is obtained by equating the coefficients of (z^{-2}), i.e., $N2$ and $D2$, to zero.

The designer must decide how to pair the poles and zeros in forming the $D_i(z)$. A pole-zero pairing algorithm that minimizes the output noise is available. The ordering of the $D_i(z)$ also affects output noise due to quantization and whether or not limit cycles are present.

For the parallel realization (see Fig. 7.3), the transfer function is expanded as a sum of first-order and second-order expressions of the form:

$$D(z) = K + D_1(z) + D_2(z) + \cdots + D_n(z) \tag{7.18}$$

where the first-order blocks have the form:

$$D_i(z) = \frac{NO}{1 + D1z^{-1}} \tag{7.19}$$

and the second-order blocks have the form:

$$D_i(z) = \frac{NO + N1z^{-1}}{1 + D1z^{-1} + D2z^{-2}} \tag{7.20}$$

The concept of ordering of $D_i(z)$ does not apply to the parallel configuration. Pole-zero pairing is fixed by the constraints imposed by the partial-fraction expansion.

The parallel realization has an obvious advantage over the cascade form

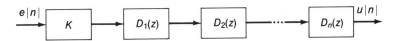

Figure 7.2   Cascade implementation of a high-order transfer function

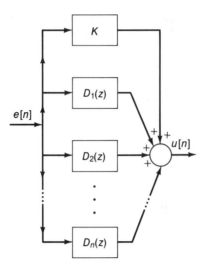

Figure 7.3 Parallel implementation of a high-order transfer function

when an algorithm executes on a multiprocessor system, where the filter algorithm can be split up among the processors and run concurrently.

No significant difference is apparent between the parallel and cascade realizations in performance factors, such as execution speed and program/data memory use, when the algorithms execute on a single processor. The parallel algorithm could possibly provide more precision in computing the filter's output, if the designer decided to save the double-precision (32-bit) results from each of the first- and second-order sections and perform a double-precision addition to calculate the final filter output.

TMS 320 Example Program

An example TMS 32010 program that uses the first- and second-order filter routine to compute several elements of a digital control system is provided in this section. The program illustrates the concepts of modular software design, data memory layout, and cascade implementation of high-order transfer functions. The program was executed on a combination of the TMS 32010 Emulator (XDS) and Analog Interface Board (ATB) using random noise as input. The input was sampled at a 4000-Hz rate.

The following transfer functions are implemented with Q12 scaling:

$$900 - \text{Hz Notch Filter } D(z) = \frac{0.8352 - 0.2729z^{-1} + 0.8352z^{-2}}{1.0 - 0.2729z^{-1} + 0.60704z^{-2}}$$

$$1800 - \text{Hz Notch Filter } D(z) = \frac{0.9688 + 1.8341z^{-1} + 0.9688z^{-2}}{1.0 + 1.8341z^{-1} + 0.9375z^{-2}}$$

Other transfer functions (compensators, notch filters) can be implemented

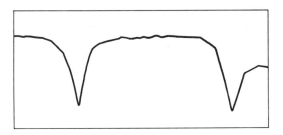

Figure 7.4 Spectrum analyzer output (900-Hz and 1800-Hz notch filters)

in identical fashion by expanding the data structure (filter coefficients and states) and making additional filter routine calls to compute these elements. The output, as observed on a spectrum analyzer, is shown in Fig. 7.4. The first notch from the left is at 900 Hz, the second is at 1800 Hz. The attentuation of the notch frequencies is about 23 dB, with respect to the passband region.

7.5 PROCESSOR INTERFACE CONSIDERATIONS

Alternatives should be considered when designing the data-acquisition portion of the digital controller hardware. This section addresses the A/D and D/A converter selection, different analog sensor interface methods, and communication with the host processor.

A/D and D/A Conversions and Integrated Circuits

The A/D and D/A converter selection for a control system design may be based on several factors. Among the most crucial factors are the maximum conversion speed of the converter and the wordlength of the device.

The A/D conversion speed relates directly to the required sampling rate of the specific application. This rate is determined by the need to sample fast enough to prevent aliasing and excessive phase lag, and to sample slow enough to avoid the unnecessary expense and accuracy of high data rates.

The A/D wordlength should be chosen based on a worst-case analysis using the following two criteria:

1. The dynamic range of the continuous input signal
2. The quantization noise of the A/D converter

For dynamic range, the designer should determine the minimum and maximum values of the continuous input that needs to be accurately represented and select the A/D wordlength in bits based on the resolution required within this range.

Quantization noise is due to the quantization effect of the A/D. The value of this noise during a single conversion can be represented by the difference between the exact analog value and the value allowable with the finite resolution of the A/D. This quantization noise may assume any value in the range $-q/2$ to $+q/2$, for a rounding converter, or 0 to q for a truncating A/D converter, where q is the quantization level. The quantization noise may assume any value in the range $-q/2$ to $+q/2$, for a rounding converter, or 0 to q for a truncating A/D converter, where q is the quantization level. The quantization level q is equal to the full-scale voltage range, divided by 2^B, where B is the number of bits in the converter. The quantization noise may be modeled as uniformly distributed noise. The designer should make the choice based on the maximum acceptable quantization level.

The D/A converter wordlength should be chosen in a similar manner to choosing the A/D wordlength, by considering the dynamic range of the output signal.

The effects of A/D and D/A converter wordlength on the performance of a high-speed control system are detailed in the University of Arkansas study (see the section on the design example of the TMS 32010-based rate-integrating gyro positioning system). The study analyzed the time-domain performance of the system (unit impulse, step, ramp, and torque-disturbance response) as a function of A/D and D/A wordlengths. Twelve-, 14-, and 16-bit converters were used. The only significant difference found between them was the steady-state error. Twelve-bit converters were found to be adequate.

In a multi-input digital control system, the signal acquisition portion of the digital controller must provide for the multiplexing of several analog inputs into a single A/D converter. Consequently, some external devices are needed to prefilter (antialiasing filters), sample, and hold the analog signals from each channel (S/H circuits), and multiplex the signals onto the A/D converter (analog multiplexer). Multiplexing and filtering may also be necessary at the output, in cases where the digital control system computes multiple outputs for the control of the plant.

Two configurations of a cost-effective, multichannel data acquisition system for a digital controller are shown in Fig. 7.5. The first accommodates up to eight inputs; the second can accommodate up to 32 inputs. Note that in these two systems, only one S/H per eight inputs exists, and the variables are sampled in sequence with the same sampling interval between successive samples of a given signal. There will be a "skew" in time between the samples of the various inputs with a possible unwanted effect on the system performance. In this case, the use of a fast A/D converter may be justified to minimize this effect.

If truly simultaneous sampling is required, an array of S/H circuits may be used to capture the values of all the inputs concurrently. This solution, shown in Fig. 7.6, is more expensive due to the cost of S/Hs. In such simultaneous sampling systems, a fast conversion must be performed before the signal values present on the S/Hs start to droop. Therefore, the maximum conversion rate must be fast enough to accommodate this constraint.

In some cases, high-speed A/D converters (100- to 500-kHz conversion rates)

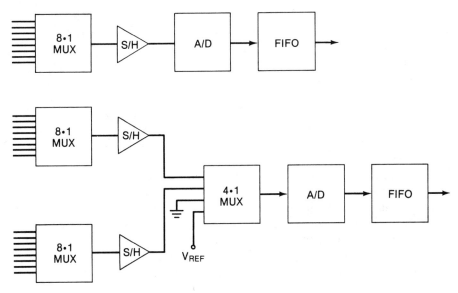

Figure 7.5   Cost-effective data acquisition systems

are required. Two 12-bit A/D devices that will accommodate these speed requirements are the ADC85 and the AD5240 from Analog Devices. The ADC85 allows
a conversion rate of up to 100 kHz and AD5240 up to 200 kHz. There are other
converters available from several manufacturers. D/A converters, on the other
hand, are inherently faster, the selection is much broader, and the cost is less.

When high-speed converters are not needed (when only a few input channels
exist or a single converter per channel is justified), devices such as the TCM 2913
and TCM 2914 codecs may be useful in digital control applications. Although
telecommunications-oriented, these devices are low in cost and provide on-chip

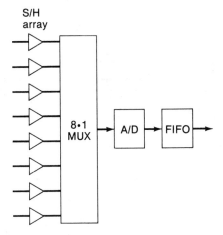

Figure 7.6   Data acquisition system with
simultaneous sampling

antialiasing and smoothing filters. The TCM 2913 or TCM 2914 both contain A/D and D/A converters. The 8-bit digital output of the A/D and the 8-bit digital input to the D/A are both arranged in a companded (expanded/compressed) form, using -law or A-law companding techniques. The -law and A-law companding techniques allow small numbers to be represented with maximum accuracy, but require a conversion routine before the commanded samples can be used in two's-complement computations. Such conversion routines are based on lookup tables and need only a few TMS 32010 instruction cycles to execute. The devices interface to the processor in a serial form and convert the data at a maximum rate of 8 kHz.

All of these data acquisition systems can accommodate differential inputs from analog transducers, such as pressure sensors, strain gauges, and others. To maintain accuracy in the case of a low-level input signal and to minimize noise effects, twisted-pair leads can be connected to a differential analog multiplexer, which drives an instrumentation amplifier of the same kind. The amplifier rejects the common-mode noise and presents the single-ended output to the S/H circuit and A/D converter. These two configurations are shown in Fig. 7.7.

Sychronization of the Processor and External Devices

In a multichannel data acquisition system with one A/D converter, a designer must generate a sequence of timing signals to synchronize the S/H circuits, the analog multiplexer, the A/D converter, and the input/output latches with the operation of the TMS 32010. The designer may decide to generate the timing signals from the TMS 32010 clock, by subdividing its frequency or using a timing and control circuit based on its own clock.

An alternative way to build a multichannel data acquisition system is to designate a separate A/D for each channel. In this case, the timing-signal generation is simpler and the A/D converters used may be slower and less costly,

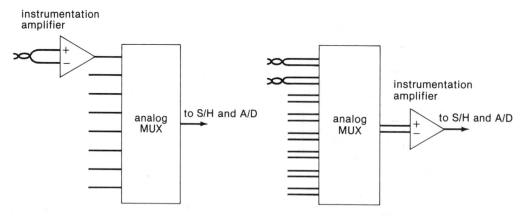

Figure 7.7   Differential input configurations

although more of them are necessary. The designer should perform a tradeoff analysis based on board space, overall system cost, and power consumption.

Communications with Host Computer

In addition to having a fast signal-processing microprocessor, a need may exist for an executive processor to monitor the system's operation. Such an executive processor would be used for system startup/initialization (coefficient and initial-condition loading), responding to emergency conditions, such as overflow and underflow, system reprogramming/reconfigurating (loading a new program or a new set of coefficients), and thorough system test and calibration. The system should be constructed so that the executive processor can interrupt, halt, or alter the execution of the signal processor at any time in response to contingency situations.

BIBLIOGRAPHY

[7.1] C. Slivinsky and J. Borminski, ''Control System Compensation and Implementation with the TMS 32010,'' Digital Signal Processing, Application Report, Texas Instruments Co. Ltd., U.S.A.

Chapter 8

Control Program
Development Procedures
Using a TMS 320

8.0 INTRODUCTION

A servo motor control program written by assembly language is explained in this chapter. An example, PI (proportional and integral) servo motor speed controller, is used for explanation. However, this control program can be utilized for writing another control algorithm described in chapters 5 and 6 by just replacing the following control calculation part.

The flowchart for this PI control program is shown in Fig. 8.1. The entire control program is listed in Fig. 8.2. The following is a part by part explanation of the whole program: The system shown in Fig. 1.14 is used.

8.1 CONSTANTS AND VARIABLES DECLARATION
AND SUBSTITUTION

```
RESLT EQU >4000
```

> shows that 4000 is a hexadecimal number.
The hexadecimal number 400 is substituted into RESLT. It is also possible to

178

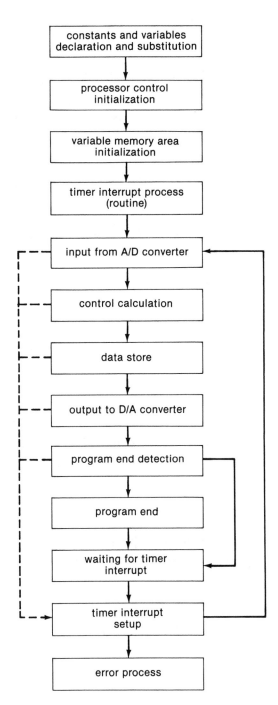

Figure 8.1 Flowchart for PI control program

```
0
      * * * * * * * * * * * * * * * * * * * * * * * * * * * * * * * * * * * * * * * * * * * * * * * * * * *
      *        1987/05/7                                                          *
      *                                                                          *
      *                     ASSEMBLER PROGRAM FOR SERVO MOTOR                      *
      *                                                                          *
      *                                                                          *
      * * * * * * * * * * * * * * * * * * * * * * * * * * * * * * * * * * * * * * * * * * * * * * * * * * *
      *
      *        Address
      *
RESLT   EQU       >4000        *data store address
      *
      *        2 port RAM
      *
ERRMSG  EQU       0            *error message address
ENDMSG  EQU       1            *end    message address
      *
      *        VALUE
      *
        DORG      0
      *
*constants
*
SAMPTM  BSS       1            *sampling time
DCOUNT  BSS       1            *driving counter
GOAL    BSS       1            *reference value
PGAIN   BSS       1            *p gain
IGAIN   BSS       1            *i gain
OFFSET  BSS       1            *offset of A/D converter1
*
*variables
*
WRKST   EQU       $
*
RECPTR  BSS       1            *record pointer
TIMEFG  BSS       1            *timer interrupt flag
OMEGA   BSS       1            *angle velocity of motor
MODEL   BSS       1            *reference model
ERR     BSS       1            *error
INTEG   BSS       1            *integral value
DINTEG  BSS       1            *decimal of integral value
CNTRL   BSS       1            *output control value
REG1    BSS       1            *
REG2    BSS       1            *
```

Figure 8.2   Control program

```
REG3      BSS      1          *
*
WRKEND    EQU      $-1
*
 ***************************************************************
 *                                                             *
 *        INITIALIZATION                                       *
 *                                                             *
 ***************************************************************
          AORG     >2000
*         processor initialization
          DINT                        *disable interrupt
          SOVM                        *set overflow mode
          SSXM                        *set sign-extension mode
          SPM      0                  *set shift of p register 0
          CNFD                        *configure block B0 as data
                                       memory
*
          LDPK     4                  *DP<--4
          LARP     AR0                *ARP<--0
*
*         work area initialization
*
          LALK     WRKEND             *reg1<--size of work area
          SBLK     WRKST
          SACL     REG1
*
          LALK     WRKST              *AR0<-- >200+wrkst
          ADLK     >200
          SACL     REG2
          LAR      AR0,REG2
*
          ZAC                         *ACC<--0
          RPT      REG1               *repeat (wrkend-wrkst+1)
                                       times
          SACL     *+                 *DM(AR0)<--0 : AR0<--AR0+1
*
          LALK     RESLT              *recptr<--reslt
          SACL     RECPTR
*         servomotor initialization
          OUT      REG1,PA0           *motor<--0 volt
*
*         timer initialization
*
          LALK     TCHECK             *reg1<--address of timer check
                                       routine
```

Figure 8.2 (*continued*)

```
          SACL      REG1
          LALK      25                  *ACC<--25
          TBLW      REG1                *25(PM) --reg1
*
          LAC       SAMPTM
          LDPK      0                   *DP<--0
          SACL      2                   *2(TIM)<--sampling time
          SACL      3                   *3(PRD)<--
*
          LAC       4                   *mask 4(IMR)
          ORK       >0008
          SACL      4                   *enable timer interrupt
*
          LDPK      4                   *DP<--4
          EINT                          *enable interrupt
*
          IDLE                          *idle until interrupt
*****************************************************************
*                                                               *
*         MAIN ROUTINE                                          *
*                                                               *
*****************************************************************
MAIN      EINT                          *enable interrupt
*
*         input from A/D converter
*
          IN        OMEGA,PA0           *omega<--input
          LAC       OMEGA
          ADD       OFFSET              *compensate offset of D/A
                                        converter
          SACL      OMEGA
*
*         make model (reference)
*
          LAC       MODEL
          SUB       GOAL
          BGEZ      CONST               *if model>=goal then go to
                                        const
          LAC       MODEL               *model<--model+8
          ADLK      8
          SACL      MODEL
*
*         control calculation
*
```

Figure 8.2 (*continued*)

```
CONST    LAC      MODEL          *err=model-omega
         SUB      OMEGA
         SACL     ERR
*
         LT       ERR
         MPY      IGAIN          *(16.16)=err(16.0)
                                 *igain(0.16)
         PAC                     *i control(integ)=err*igain
         ADDH     INTEG          *                      +integ
         ADDS     DINTEG         *                      +dinteg
*
         SACH     INTEG
         SACL     DINTEG
*
         MPY      PGAIN          *p control(32.0)=err(16.0)
                                 *pgain(16.0)
         PAC
         ADD      INTEG          *control=i control+p control
         SACL     CNTRL
*
*        output to D/A converter
         SBLK     >7FFF
         BLZ      MIN            *if control>(>7FFF) then
                                 control=(>7FFF)
         LALK     >7FFF
         SACL     CNTRL
         B        OUTPUT
*
MIN      ADLK     >7FFF,1
         BGEZ     OUTPUT         *if control<(>8000) then
                                 control=(>8000)
         ADLK     >8000
         SACL     CNTRL
*
OUTPUT   OUT      CNTRL,PA0      *output<--control
*
*  experimental data record
*
         LAR      AR0,RECPTR     AR0<--record pointer
*
         LAC      OMEGA
         SACL     *+             *store omega : AR0=AR0+1
         LAC      MODEL
         SACL     *+             *store model : AR0=AR0+1
         LAC      CNTRL
         SACL     *+             *store cntrl : AR0=AR0+1
```

Figure 8.2 (*continued*)

```
        LAC       INTEG
        SACL      *+                    *store integ : AR0=AR0+1
*
        SAR       AR0,RECPTR            *record pointer<--AR0
*  program end detection
        LAC       DCOUNT                *dcount=dcount-1
        SBLK      1
        SACL      DCOUNT
        BNZ       WATE                  *if dcount=0 then program end
***********************************************************************
*                                                                     *
*                     PROGRAM END                                     *
*                                                                     *
***********************************************************************
        DINT                            *disable interrupt
*
        ZAC
        SACL      REG1                  *motor 0 volt
        OUT       REG1,PA0
*
        LDPK      0                     *DP<--0
        LAC       4                     *mask IMR
        ANDK      >FFF7
        SACL      4                     *disable timer interrupt
*
        LDPK      32                    *DP<--32 page
        LALK      >FFFF
        SACL      ENDMSG                *send 9801 end message
*
        RET                             *return to IPL program
***********************************************************************
*                                                                     *
*                 WAITING FOR INTERRUPT                               *
*                                                                     *
***********************************************************************
WAIT    ZAC                             *timefg<--0
        SACL      TIMEFG
        IDLE                            *idle until interrupt
        B         MAIN                  *goto main
***********************************************************************
*                                                                     *
*                 TIMER INTERRUPT ROUTINE                             *
*                                                                     *
***********************************************************************
TCHECK  LAC       TIMEFG
        BNZ       TMERR                 *if timefg<>0 then goto tmerr
```

Figure 8.2 (*continued*)

```
        LALK    >FFFF               *timefg<--FFFF
        SACL    TIMEFG
*
        RET
*
*       timer interrupt err routine
*
TMERR   DINT                        *disable interrupt
*
        ZAC
        SACL    REG1                *motor<--0 volt
        OUT     REG1,PA0
*
        LDPK    0                   *DP<--0
        LAC     4                   *mask IMR
        ANDK    >FFF7
        SACL    4                   disable timet interrupt
*
        LDPK    32                  *DP<--32
        LALK    >FFFF
        SACL    ERRMSG              *send 9801 err message
*
        POP                         *pop stack register
        RET                         *return to IPL program
```

Figure 8.2   (*continued*)

express it by using a decimal number 16384 as follows:

```
RESLT EQU 16384
DORG    0
```

The addresses for the following constants and variables are ones assigned from the address number 0.

The constants and variables are defined as:

RESLT: The initial memory address for a block of measured data storage memories.

ERRMSG: This is the address on the 2 port RAM (page on data memory), which is used for informing the host computer (NEC PC-9801) of an error in the timer interrupt.

ENDMSG: This is also the address on the 2 port RAM, which is utilized for passing the message "program end" to the host computer.

SAMPTM BSS 1

One memory address is assigned for the constant SAMPTM. Also, multiple memory addresses can be assigned as follows:

```
DORG        0
SAMPTM BSS 1
DCOUNT BSS 1
```

Equivalently:

```
SAMPTM EQU 0
DCOUNT EQU 1
WRkST EQU   $
```

The following assigned memory address is substituted into (is equivalent to):

```
WRKST:
      WRKST = address of RECPTR
WRKEND EQU $-1
```

The address assigned just one before this instruction is equated to WRKEND.

```
WRKEND = address of REG 3
```

The contents in WRKEND and WRKST are useful for the initialization of the variables memory area.

The addresses for the constants and the variables are defined as follows:

SAMPTM	This is the sampling period (timer interrupt period) in the control calculation. The period is that of SAMPTM × 0.8 µs.
DCOUNT	The number of control calculations. When the number of control calculations is equivalent to the content in DCOUNT, the program ends.
GOAL	The motor speed reference GOAL = 100 H (500 rpm).
PGAIN	The proportional controller gain
IGAIN	The integral controller gain
OFFSET	The offset compensation value for the A/D converter
RECPTR	The memory address in which the measured data are stored
TIMEFG	The flag for showing the error in the timer interrupt generation
OMEGA	The servo motor speed
MODEL	The servo motor speed reference (desired speed)
ERR	The error (MODEL-OMEGA)
INTEG	The integer part of the integral controller output
DINTEG	The fractional number part of the integral controller output

CNTRL   The control to the servo motor drive system
REG1(2,3)   The registers used for things such as storing tentative control calculation results

The variables and constants defined from SAMPTM to REG 3 are stored on page 4 of the data memory.

8.2 INITIALIZATION

Program Initialization

A or G >2000

>2000 is assigned as the initial program memory address for storing the following program.

Processor Initialization

The mode of operation, the function of the memory pointer, and others are explained as follows:

DINT	interrupt prohibition (except when resetting)
SOVM	setting the overflow mode
SSXM	setting the code extension mode
SPM	0 prohibits shift in the output of the P register
CNFD	the block BD is used for data memory
LDPK	4 the page pointer is set at 4
LARP ARO	the auxiliary pointer is set to ARO

Work Over Initialization

In this part, the variables' memories (on the address number 6 through 16, page 4) in the program are filled with zeros for their initialization. The initial memory address WRKST is substituted into ARO, then zeros are stored in all memories by indirect addressing and the contents of ARO is increased by one. This procedure is repeated (WRKEND − WRKST + 1) times.

Timer Initialization

An inner timer interrupt is introduced in order to generate control at a constant sampling rate. The following three memory map registers and interrupt vector locations are required:

Memory map register (on page 0 of the data memory)

By placing one in the 4th most insignificant bit of the IMR, a timer interrupt is executed. The following explains the mechanism of the timer interrupt.

TIM is counted down (decremented by one) every $0.8s$

When the contents of TIM becomes zero, a timer interrupt takes place. The program (control) branches to the routine, defined in the interrupt vector locations, when the timer interrupt is generated, the content of the PRD is loaded into TIM during one period. This results in generation of the constant period timer interrupt.

8.3 MAIN ROUTINE

Main EINT

This instruction enables the interrupt. In the program, the interrupt routine address TECECK is stored in address #25 of program memory. Thus, when the timer interrupt is generated, the program branches to the label of TECH. The sampling rate is stored in TIM and PRD. In order to set the 4th most insignificant bit of IMR, logical OR operation of IMR and 0008H is taken and the result is placed in IMR. After the timer interrupt is enabled, it is permitted and then the program returns to the main routine. While an interrupt is in execution, the net interrupt is prohibited. Therefore, when a new interrupt generation is required, the instruction EINT is put again.

Input from A/D Converter

The output of the A/D converter through the address #0 on the I/O port is obtained for the motor speed OMEGA, then the offset of the A/D converter is compensated.

Control Calculation

The following servo motor speed reference signal, shown in Fig. 8.3, is generated in the "make model (reference)."

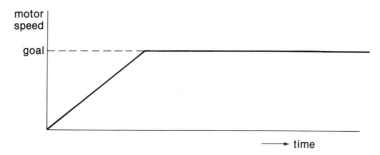

Figure 8.3   Reference model

A well-known PI (Proportional and Integral) control algorithm is used in this program. The control block diagram is shown in Fig. 8.4. The error e is the difference between the speed command (reference) and measured servo motor speed as measured by tachometer. The PI feedback control loop is utilized to nullify this error.

Proportional control is the product of the error and proportional gain K_p. The integral control is the product of the error and the integral gain K_I. K_I takes a small value. In the program, K_I is expressed by using 16 bits for its fractional number. For the integer part of K_I, 16 bits are also utilized. Therefore, the integer part and fractional part of K_I are stored individually in different memories. The least significant 16 bits are summed (integrated) by using ADDSS without the code extension. The total control is the sum of the P control and the I control.

Output to D/A Converter

If the calculated control overflows beyond 2^{16}, then the control takes the maximum value or negative minimum value. The control passes to the D/A converter through the address #0 of the I/O port.

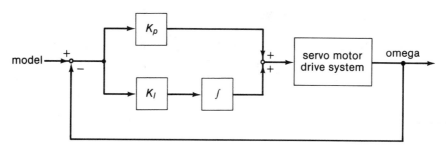

Figure 8.4   PI control block diagram

Experimental Data Record

OMEGA (motor speed), MODEL (motor speed command), CNTRL (control), and INTEG (integral control) are stored by using indirect addressing from the address #RESLT. The number of memories that have been used is found by judging from the content of RECPTR (the last memory address).

Program End Detection

The content of DCOUNT is decremented once for each control calculation. If the content is zero, the program ends.

8.4 PROGRAM END

The interrupt is disabled. The servo motor is at the stall position. IMR is reset. The program end message is sent to the host computer through 2 port RAM. Finally, the program returns to the IPL (Initial Program Loading) program.

8.5 WAITING FOR INTERRUPT

If the control calculation is finished within the sampling period, TIMEFG is set at zero. There is a wait for the timer interrupt; then, when the program returns from the timer interrupt routine, it jumps to the main routine.

8.6 TIMER INTERRUPT ROUTINE

If TIMEFG is not equal to zero, the control calculation is not finished within the sampling interval. This is processed as the error is generated. If TIMEFG is zero, then the program returns to the main routine after substituting FFFFH into TIMEFG.

Timer Interrupt Error Routine

In this program the interrupt is disabled, the motor rotation is stopped, and IMR is reset. Then, the error message passes to the host computer through the 2 port RAM. The stack register is topped up and execution returns to the IPL program. If there is the instruction RE instead of POP, then execution returns to the main routine.

Chapter 9

Implementation of Position, Speed, and Force Controls with a TMS 320

9.0 INTRODUCTION

Three examples of using the TMS 32020 processor to implement digital position, speed, and force control systems are presented in this chapter. The system's hardware and software and system performance are discussed.

9.1 SYSTEM DESCRIPTIONS

The systems used as examples of the application of the TMS 320C25 are the servo motor control systems for trajectory (position) tracking control, for a two-degrees of freedom manipulator with a motor drive, speed regulation, and tracking control, for a brushless servo motor and gripping force control for a manipulator hand with a DC-motor drive. Such systems are required for the precise (robust) and rapid position, velocity, and force control (motion control) with servo motor drives. Figures 9.1, 9.2, and 9.3 show the configurations of the two-degrees of freedom manipulator, the brushless servo motor, and the manipulator hand, respectively.

At present, digital control with DSPs is not normally used in these types of systems; however, more robust and much faster control can be expected, due to the faster throughput rates and higher computational accuracy of the DSP.

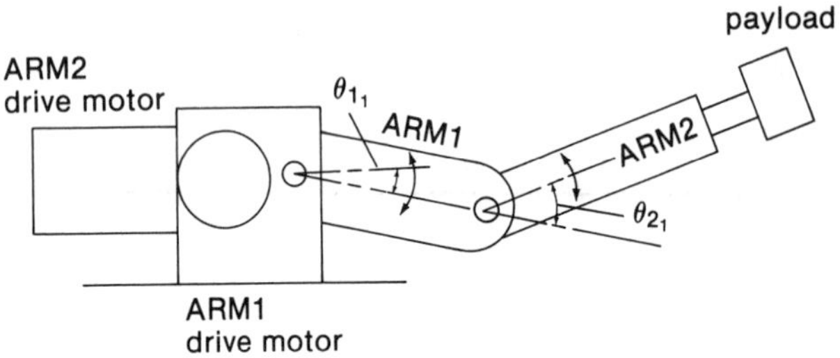

Figure 9.1 Configuration of robot manipulator. DC servo motor with reduction gear ratio 1/50 and arm drive gear ratio 1/4.

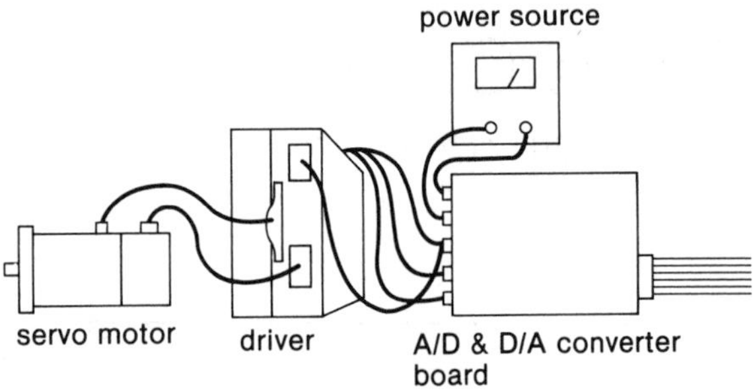

Figure 9.2 Brushless servo motor drive system configuration

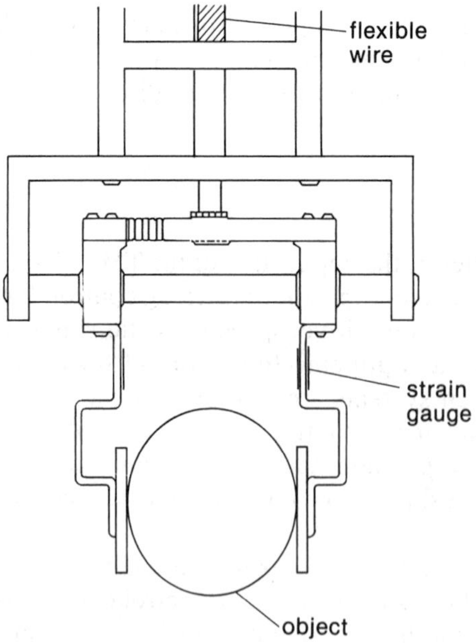

Figure 9.3 Manipulator hand configuration

9.2 CONTROL REALIZATION (BLOCK DIAGRAMS AND FLOWCHARTS)

In this section, control block diagrams, flowcharts, and control programs written by the TMS 320 assembly language for position, velocity, and force control systems are given.

Control Block Diagrams

Control block diagrams based on the control algorithms developed in Chapter 5 are illustrated. Figure 9.4 shows the control block diagram for position (trajectory) tracking of a two-degrees of freedom manipulator. Adaptive gain feedforward control, based on a sliding curve and variable structure PI control, whose deviation is based on sliding mode control theory, are used. The input reference (desired trajectory) is generated in the program. The following constants and variables are utilized:

$$\phi_0 = \left(1 + \frac{s\ell}{|s\ell| + \delta_f} \frac{\dot{g}}{|\dot{g}|} \right) K_f,$$

$$s\ell i = \dot{e} + ce$$

$$Kv_p = \frac{s\ell}{|s\ell| + \delta_b} K_p,$$

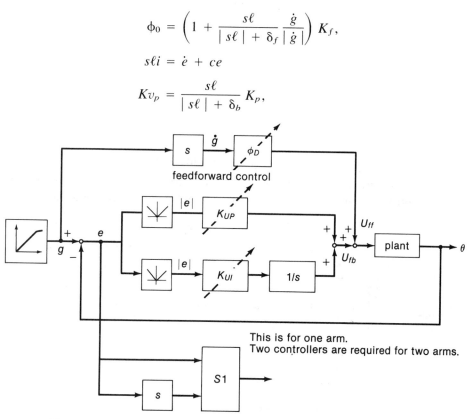

Figure 9.4   Control block diagram for trajectory position tracking of two-degrees-of-freedom manipulator

$$Kv_I = \left(1 - \frac{|s\ell|}{|s\ell| + \delta_b}\right) K_I$$

$$K_f = 4864 \ (1300 \ H)$$

$$\delta_f = 1., \ K_p = 4096 \ (1000 \ H),$$

$$\delta_b = 1., \ k_I = 4.$$

sampling time: 100 μsec $C = \frac{1}{16}$

Figure 9.5 shows the control block diagram for brushless servo motor speed regulation and tracking. Variable structure PI control is applied and the desired model is used. The following constants and variables are utilized:

$$Kv_p = \frac{s\ell}{|s\ell| \delta} K_p$$

$$Kv_p = \left(1 - \frac{|s\ell|}{|s\ell| + \delta}\right) K_I,$$

$$s\ell = \dot{e} + ce$$

$$K_p = 256$$

$$K_I = \frac{1}{32}$$

$$C = \frac{1}{64}$$

$$\delta = 1,$$

sampling time 50 μsec

Figure 9.6 shows the block diagram for a manipulator hand gripping force control system. Fuzzy set control and improved sliding mode control are used.

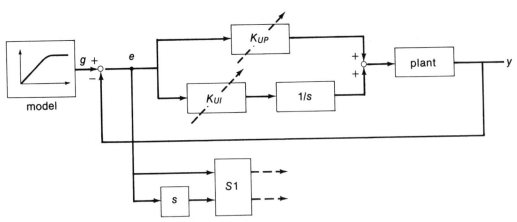

Figure 9.5   Control block diagram for brushless servo motor speed regulation and tracking

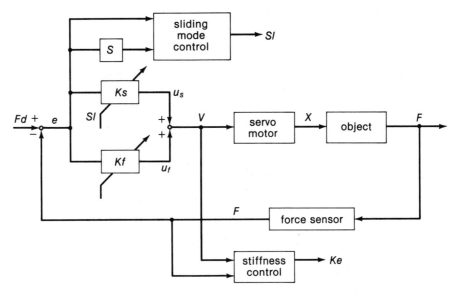

Figure 9.6 Control block diagram for manipulator hand gripping force control

The constants and variables used here are as follows:

$$U_s = \frac{s\ell}{|s\ell| + \delta_s} K_s |e(k)|,$$
$$u_f = K_f e(K)$$

$$K_f = C_{s,f} N / \left\{ \sum_{i=0}^{N} K_e(k - i) \right\}$$

$$s\ell = ce(k) + \dot{e}(k)$$

$$K_e(k) = \left\{ \begin{array}{l} K_{e\max}(K_e > K_{e\max}) \\ K_{e\min}(K_e < K_{e\min}) \end{array} \right\}$$

$$e(k) = F_d - F(k)$$

$$\{F(k) - F(k - 1) + \delta_{frc}\}\{1/v_{(k-j)} + \delta\}$$

$$\dot{e}(k) = e(k) - e(k - 1)$$

$$(K_{e\min} \le K_e \le K_{e\max})$$

A moving average filter is used here.

$$V = U_s + U_f = K_p e(k)$$

$$K_p = \frac{s\ell}{|s\ell| + \delta_s} \frac{e(k)}{|e(k)|} K_s + K_f$$

$F_d = 2048$

$K_{f\min} = 6.51898,$

$K_s = 3$

$N = 10,$

$K_f \times K_e = 0.0189991$

$K_{f\max} = 1.40059$

$\delta_s = 30,$ $\delta_{fre} = 0.0625$ $\delta_{1/N} = 0.0012207$

Sampling time: 0.8 msec

The constants used here have been determined by experiments.

Flowcharts and Program Lists

Flowcharts for explaining the control programs corresponding to these three examples are shown in Figs. 9.7, 9.8, and 9.9, respectively. The variables and constants used in the flowcharts are described in the programs.

Control programs written in the TMS 320 assembly language corresponding to these examples are listed in Figs. 9.10, 9.11, and 9.12, respectively. The explanations in details for the programs are also found in Chapter 8.

9.3 HARDWARE IMPLEMENTATION

TMS 320 Memory Map

Hardware configurations corresponding to three examples are illustrated in Figs. 9.13, 9.14, and 9.15, respectively. A digital interface board between a host computer and a TMS 320C25 is commercially available. Figure 9.16 shows the TMS 320 memory map.

Servo Motors, Drivers, and Sensors

The following DC servo motors and drivers are used for position tracking and gripping force controls:

Yasukawa's Printer Motor PMS-09AFF

rated output = 100 W   rated voltage = 26 V

rated torque = 2.43 Kgf-cm   rated current = 5.5 A

rated max.

rpm = 4000 rpm   power rate = 1.3 kW/sec

mechanical time constant = 10 msec

electrical time constant = 0.04 msec

Yasukawa's servo driver *Servo Pack CPCR-FR01AB*

input power = 300 VA, 200/220 VAC 50/60 Hz

max. output voltage ± 30 V (± 6 A)

max. output current ± 6 A

speed control ratio = 1:1000

input voltage = $\pm(10)$ V

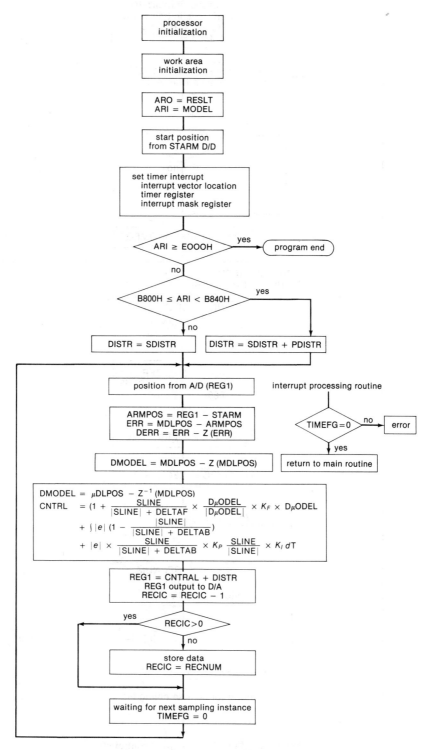

Figure 9.7 Flowchart for position tracking control program

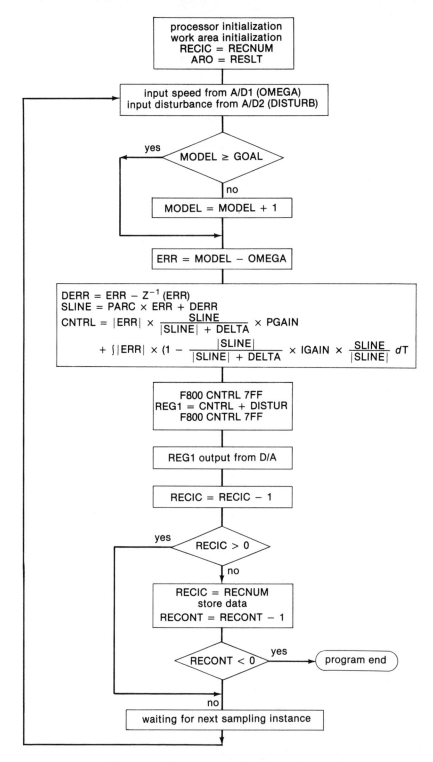

Figure 9.8 Flowchart for speed control program

198

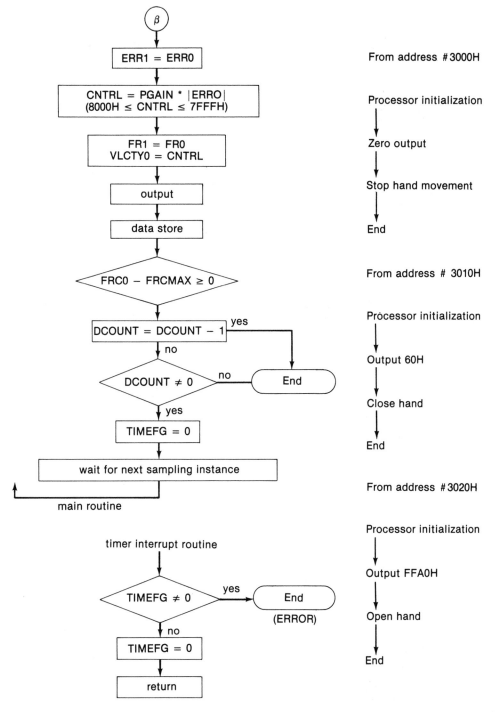

Figure 9.9 Flowchart for gripping force control program

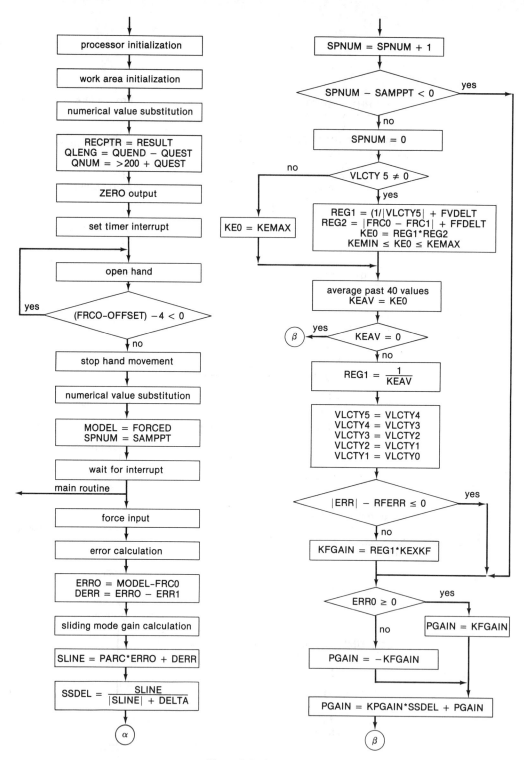

Figure 9.9 (*continued*)

```
***********************************************************************
*       1987/07/25                                                    *
*                                                                     *
*                  ASSEMBLER PROGRAM FOR MANIPULATOR                  *
*                                                                     *
*                                                                     *
***********************************************************************
*
*          Address
*
RESLT    EQU      >4000       *data store address
MODEL    EQU      >A000       *model address
*
*          2 port RAM
*
ERRMSG   EQU      0           *error message address
ENDMSG   EQU      1           *end     message address
*
*          VALUE
*
         DORG     0
*
*constants
*
SAMPTM   BSS      1           *sampling time
RECINT   BSS      1           *record's interpose
PARC     BSS      1           *c of switching line
DELTAF   BSS      1           *delta f
DELTAB   BSS      1           *delta b
PGAIN    BSS      1           *p gain of feedback
IGAIN    BSS      1           *i gain of feedback
FGAIN    BSS      1           *feedfoward gain
PDISTR   BSS      1           *pulse disturbance
SDISTR   BSS      1           *steady disturbance
*
*variables
*
WRKST    EQU      $
*
STARM    BSS      1           *start arm position
RECPTR   BSS      1           *record pointer
RECIC    BSS      1           *record interpose counter
TIMEFG   BSS      1           *timer interrupt flag
ARMPOS   BSS      1           *arm position
MDLPOS   BSS      1           *model
DMODEL   BSS      1           *d/dt(model)
```

Figure 9.10 Manipulator control program

```
ERR       BSS       1              *error
DERR      BSS       1              *d/dt(err)
YSTERR    BSS       1              *1/z(err)
SLINE     BSS       1              *switching function
INTEG     BSS       1              *integral value
DINTEG    BSS       1              *decimal of integral value
PCNTRL    BSS       1              *proportional control of feedback
ICNTRL    BSS       1              *integral control of feedback
FCNTRL    BSS       1              *feedforward control
CNTRL     BSS       1              *output control value
DISTR     BSS       1              *total disturbance
REG1      BSS       1              *
REG2      BSS       1              *
REG3      BSS       1              *
REG4      BSS       1              *
WRKEND    EQU       $-1
*

*************************************************************
*                                                           *
*         INITIALIZE                                        *
*                                                           *
*************************************************************

          AORG      >2000
*
          DINT                     *disable interrupt
          SOVM                     *set overflow mode
          SSXM                     *set sign-extension mode
          SPM       0              *set shift of p register 0
          CNFD                     *configure block B0 as data
                                    memory
*
          LDPK      4              *DP<--4
          LARP      AR0            *ARP<--0
*
          ZAC                      *ACC<--0
          SACL      REG1           *motor 0 volt
          OUT       REG1,PA9
*
*         fill zero at work area
*
          LALK      WRKST
          ADLK      >200
          SACL      REG1
          LRLK      AR0,REG1       *AR0<--start address of work
                                    area
          LALK      WRKEND         *reg1<--size of work area
```

Figure 9.10 (*continued*)

```
        SBLK      WRKST
        SACL      REG1
*
        ZAC                           *ACC<--0
        RPT       REG1                *fill from wrkst to wrkend with
                                      0
        SACL      *+
*
*
*
        LRLK      AR0,RESLT           *AR0<--reslt
        LRLK      AR1,MODEL           *AR1<--model
*
        LALK      >FFFF               *counter latch
        SACL      REG1
        OUT       REG1,PA10
        IN        STARM,PA9           *store start arm position
*
*       timer initialize
*
        LALK      TCHECK              *25(PM)<--address of timer
                                      check routine
        SACL      REG1
        LALK      25
        TBLW      REG1
*
        LAC       SAMPTM
        LDPK      0                   *DP<--0
        SACL      2                   *2<TIM><--sampling time
        SACL      3                   *3<PRD><--
*
        LAC       4                   *mask 4<IMR>
        ORK       >0008
        SACL      4                   *enable timer interrupt
*
        LDPK      4                   *DP<--4
        EINT                          *enable interrupt
*
        IDLE                          *idle until interrupt
************************************************************
*                                                         *
*       MAIN ROUTINE                                       *
*                                                         *
************************************************************
MAIN    EINT                          *enable interrupt
*
```

Figure 9.10   (*continued*)

```
*              judgment of end
*
          RSXM                             *reset sign extent mode
          SAR        AR1,REG1              *reg1<--AR0
          LAC        REG1                  *if reg1>=E000 then ENDP
          SBLK       >E000
          BGEZ       ENDP
*
*              make disturbance
*
          LAC        SDISTR                *distr<--sdistr
          SACL       DISTR
*
          LAC        REG1                  *if B800>reg1 then INPUT
          SBLK       >B800
          BLZ        INPUT
          SBLK       >40                   *if B840<=reg1 then INPUT
          BGEZ       INPUT
*
          LAC        SDISTR                *distr<--sdistr+pdistr
          ADD        PDISTR
          SACL       DISTR
*
*              input from A/D converter
*
INPUT     SSXM                             *set code extension mode
          LALK       >FFFF                 *counter latch
          SACL       REG1
          OUT        REG1,PA10
*
          IN         REG1,PA9              *reg1<--PA0
          LAC        REG1                  *armpos=reg1-starm
          SUB        STARM
          SACL       ARMPOS
*
*              make err & d/dt(err)
*
          LARP       1                     *ARP<--1
          LAC        *                     *dmodel=mdlpos-1/z(mdlpos)
          SUB        MDLPOS
          SACL       DMODEL
*
          LAC        *+,0,0                *err=model-armpos : ARP<--0
          SACL       MDLPOS
          SUB        ARMPOS
          SACL       ERR
```

Figure 9.10   (*continued*)

```
            SUB       YSTERR           *derr=err-d/dt(err)
            SACL      DERR
*
            LAC       ERR              *ysterr=err
            SACL      YSTERR
*
*           sliding mode estimator
*
            LT        ERR              *sline(8.8)=err(16.0)
                                       *c(8.8)+derr(8.8)
            MPY       PARC
            PAC
            SACL      SLINE
*
*           adaptive gain of feedback control
*
            ABS                        *reg1=｜sl｜
            SACL      REG1
            BZ        FB0              *if ｜sl｜=0 then FB0
            ADD       DELTAB           *reg2=｜sl｜+deltab
            SACL      REG2
*
            LAC       REG1,15          *reg3(8.8)=｜sl｜/(｜sl｜+
                                       deltab)
            RPTK      8
            SUBC      REG2
            ANDK      >01FF            *mask
FB0         SACL      REG3
*
*           P feedback control
*
            LAC       ERR              *reg1=｜e｜
            ABS
            SACL      REG1
            LAC       SLINE
            BGEZ      FB1              *if sl>=0 then FB1
            LAC       REG1             *reg1=-｜e｜
            NEG
            SACL      REG1
*
FB1         LT        REG1             *reg2(24.-8)=reg1*pgain
            MPY       PGAIN
            PAC
            RPTK      7                *(24.-8)<--(16.0)
```

Figure 9.10   (*continued*)

```
        SFR
        SACL    REG2
*
        LT      REG2            *(32.0)=reg2(24.-
                                8)*reg3(8.8)

        MPY     REG3
        PAC
*
FB3     SACL    PCNTRL          *pcntrl<--ACC
        SBLK    >7FFF           *if pcntrl>7FFF then
                                pcntrl=7FFF

        BLEZ    FB4
        LALK    >7FFF
        SACL    PCNTRL
        B       FB5
FB4     ADLK    >7FFF,1         *if pcntrl+7FFF<0 then
                                pcntrl=8000

        BGEZ    FB5
        LALK    >8000
        SACL    PCNTRL
*
*       I feedback control
*
FB5     LALK    1,8             *reg2(8.8)=1-!sl!/(!sl!-
                                deltab)

        SUB     REG3
        SACL    REG2
        LT      REG1            *(24.8)=reg2(8.8)*regl(10.0)
        MPY     REG2
        PAC
        RPTK    3               *(28.4)<--(24.8)
        SFR
        BGEZ    FB6             *if ACC<0 then ACC+1
        ADLK    1
*
FB6     SACL    REG2            *reg2<--ACC
        LT      reg2            *(20.12)=reg2(12.4)
                                *igain(8.8)

        MPY     LCALK
        RPTK    3               *(16.16)<--(20.12)
        SFL
        ADDH    INTEG           *ACCH+imteg(16.0)
        ADDS    DINTEG          *ACCL+dinteg(0.16)
        SACH    INTEG           *integ=ACCH
        SACL    DINTEG          *dinteg=ACCL
        BGEZ    FB7             *if ACC<0 then ACCH+1
        ADLK    >4000,2
*
```

Figure 9.10 (*continued*)

```
FB7        SACH       ICNTRL              *icntrl<--ACCH
*
*          adaptive gain of feedforward control
*
           LAC        SLINE
           BZ         FF0                 *if sl=0 then FF0
           ABS                            *reg1=:sl:
           SACL       REG1
           ADD        DELTAF              *reg2=:sl:+deltaf
           SACL       REG2
           LAC        REG1,15             *reg2(8.8)=:sl:/
                                          (:sl:=deltaf)
           RPTK       8
           SUBC       REG2
           ANDK       >01FF               *mask
FF0        SACL       REG2
*
           LAC        SLINE               *reg2(8.8)=sl/(:sl:+deltaf)
           BGEZ       FF1
           LAC        REG2
           NEG
           SACL       REG2
*
FF1        LAC        DMODEL              *reg2=reg2*sgn(dmodel)
           BGEZ       FF2
           LAC        REG2
           NEG
           SACL       REG2
*
FF2        LAC        REG2                *reg2(8.8)=reg2+1
           ADLK       1,8
           SACL       REG2
*
           LT         REG2                *reg2(8.8)=reg2(8.8)
                                          *dmodel(16.0)
           MPY        DMODEL
           PAC
           SACL       REG2
*
           LT         REG2                *(24.8)=reg(8.8)*fgain(16.0)
           MPY        FGAIN
           PAC
           RPTK       7                   *(16.16)<--(8.8)
           SFL
           BGEZ       FF3                 *if ACC<0 then ACCH+1
           ADLK       >4000,2
*
```

Figure 9.10   (*continued*)

```
FF3      SACH     FCNTRL               *fcntrl=ACCH
         ADDH     PCNTRL               *cntrl=fcntrl+pcntrl+icntrl
         ADDH     iCNTRL
         SACH     CNTRL
         ADDH     DISTR                *reg1=cntrl+distr
         SACH     REG1
*
*        output to D/A converter
*
         OUT      REG1,PA9             *output<12bit><--control
*
         LAC      RECIC                *recic=recic-1
         SBLK     1
         SACL     RECIC
         BGZ      WAIT                 *if recic>0 then WAIT
         LAC      RECINT               *recic=recint
         SACL     RECIC
*
         LAC      ARMPOS               *store data
         SACL     *+
         LAC      MDLPOS
         SACL     *+
         LAC      DISTR
         SACL     *+
         LAC      CNTRL
         SACL     *+
*
*        waiting for sampling time
*
WAIT     ZAC                           *timefg<--0
         SACL     TIMEFG
*
         IDLE
         B        MAIN
***************************************************************
*                                                            *
*                 END OF PROGRAM
*                                                            *
***************************************************************
ENDP     DINT                          *disable interrupt
*
         ZAC
         SACL     REG1                 *motor 0 volt
         OUT      REG1,PA9
*
```

Figure 9.10   (*continued*)

```
        LDPK      0
        LAC       4
        ANDK      >FFF7
        SACL      4
*
        LDPK      32                    *DP<--32 page
        LALK      >FFFF
        SACL      ENDMSG                *send 9801 end message
*
        RET
```

Figure 9.10   (*continued*)


```
****************************************************************
*       1987/05/28                                             *
*                                                              *
*                ADAPTIVE GAIN PI SLIDING MODE                 *
*                                                              *
*                                                              *
****************************************************************
*
*           Address
*
RESLT    EQU      16384        *data store address
VALUE    EQU      4            *value page
*
*           2 port RAM
*
ENDMSG   EQU      1            *end     message address
*
*           VALUE
*
         DORG     0
*
*constants
*
RECNUM   BSS      1            *record's interpose
RECCNT   BSS      1            *drive time count
GOAL     BSS      1            *reference value
PGAIN    BSS      1            *
IGAIN    BSS      1            *
PARC     BSS      1            *
DELTA    BSS      1            *
OFFSET   BSS      1            *offset of A/D converter1
OFFST2   BSS      1            *offset of A/D converter2
*
```

Figure 9.11   Speed control program

```
*variables
*
WRKST     EQU       $
RECIC     BSS       1          *record's interpose counter
OMEGA     BSS       1          *angle velocity of motor
MODEL     BSS       1          *reference model
ERR       BSS       1          *error
YSTERR    BSS       1          *1/z(err)
DERR      BSS       1          *err-1/z(err)
INTEG     BSS       1          *integral control
INTEGD    BSS       1          *
SLINE     BSS       1          *
CNTRL     BSS       1          *output control value
DISTUR    BSS       1          *disturbance
REG1      BSS       1          *
REG2      BSS       1          *
REG3      BSS       1          *
*
WRKEND    EQU       $-1
 ************************************************************
 *                                                        *
 *         INITIALIZE                                     *
 *                                                        *
 ************************************************************
          AORG      >2000
*
          SOVM                 *set overflow mode
          SSXM                 *set sign-extension mode
          SPM       0
          CNFD                 *configure block B0 as data
                               memory
*
          LDPK      VALUE      *DP=VALUE
*
          ZAC
          SACL      REG1       *motor 0 volt
          OUT       REG1,PA0
*
*         fill zero at work area
*
          LALK      WRKST      *reg1=wrkst+200
          ADLK      >200
          SACL      REG1
          LAR       0,REG1     *AR0=reg1
          LALK      WRKEND     *reg1=wrkend-wrkst
```

Figure 9.11   (*continued*)

```
        SBLK      WRKST
        SACL      REG1
*
        LARP      0                      *ARP=0
*
        ZAC                              *fill 0 at work area
        RPT       REG1
        SACL      *+
*
        LAC       RECNUM                 *recic=recnum
        SACL      RECIC
*
        LRLK      0,RESLT                *AR0=reslt
*****************************************************************
*                                                               *
*       MAIN ROUTINE                                            *
*                                                               *
*****************************************************************
*
*       input from A/D converter
*
MAIN    IN        OMEGA,PA0              *omega <== PA0
        LAC       OMEGA                  *compensate offset of A/D
                                         converter
        ADD       OFFSET
        SACL      OMEGA
*
        IN        DISTUR,PA1             *diatur <== PA1
        LAC       DISTUR                 *compensate offset of A/D
                                         converter
        ADD       OFFST2
        SACL      DISTUR
*
*       make model
*
        LAC       MODEL
        SUB       GOAL
        BGEZ      CONST                  *if model>goal then CONST
        LAC       MODEL                  *model=model+1
        ADLK      1
        SACL      MODEL
*
CONST   LAC       MODEL                  *ysterr=model-veloci
        SUB       OMEGA
        SACL      ERR
        SUB       YSTERR                 *derr=err-ysterr
```

Figure 9.11   (*continued*)

```
          SACL    DERR
          LAC     ERR              *ysterr <== err
          SACL    YSTERR
*
          LT      ERR              *(24.8)=err(16.0)*parc(8.8)
          MPY     PARC
          PAC
          ADD     DERR,8           *S1(8.8)=err*parc+derr
          SACL    SLINE
*
          MPY     SLINE            *err*sgn(err)*sgn(S1)
          PAC
          BGEZ    EPLUS
          LAC     ERR
          NEG
          SACL    ERR
*
EPLUS     LAC     SLINE
          ABS                      *reg1 <== ¦S1¦
          SACL    REG1
          ADD     DELTA            *reg2 <== ¦S1¦+delta
          SACL    REG2
*
          LAC     REG1,15          *reg1 <==
          RPTK    8                (8.8)=¦S1¦/(¦S1¦+delta)
          SUBC    REG2
          SACL    REG1
          NEG                      *reg2 <== 1-¦S1¦/(¦S1¦+delta)
          ADLK    >100
          SACL    REG2
*
          LT      REG1             *Kvp(24.8)=Kp(16.0)*reg1(8.8)
          MPY     PGAIN
          PAC
          RPTK    7                *(16.0) <== (24.8)
          SFR
          SACL    REG1             *reg1 <== Kvp(16.0)
*
          LT      REG1             *p cntrl=err*Kvp
          MPY     ERR
          PAC
          SACL    REG1
*
          LT      REG2             *Kvi(16.16)=Ki(8.8)*reg2(8.8)
          MPY     IGAIN
          PAC
```

Figure 9.11   *(continued)*

```
         RPTK      7              *(8.8) <== (16.16)
         SFR
         SACL      REG2           *reg2 <== Kvi(8.8)
*
         LT        REG2           *(24.8)=err(16.0)*Kvi(8.8)
         MPY       ERR
         PAC
         RPTK      7              *(16.16) <== (24.8)
         SFL
         ADDH      INTEG          *err*Kvi+integ(16.0)
         ADDS      INTEGD         *         +dinteg(0.16)
         SACH      INTEG
         SACL      INTEGD
         BGEZ      IPLUS
*
IPLUS    ADDH      REG1           *reg3 <== p cntrl+i cntrl
         SACH      REG3
*
*        output to D/A converter
*
         LAC       REG3           *if reg3>7FF then reg3=7FF
         SBLK      >7FF
         BLEZ      MIN
         LALK      >7FF
         SACL      REG3
         B         CVAL
MIN      ADLK      >7FF,1         *if reg3<-7FF then reg3=F800
         BGEZ      CVAL
         LALK      >F800
         SACL      REG3
*
CVAL     LAC       REG3           *reg3=reg3+distur
         ADD       DISTUR
         SACL      REG3
*
         SBLK      >7FF           *if reg3>7FF then reg3=7FF
         BLEZ      MINUS
         LALK      >7FF
         SACL      REG3
         B         OUTPUT
MINUS    ADLK      >7FF,1         *if reg<-7FF then reg3=F800
         BGEZ      OUTPUT
         LALK      >F800
         SACL      REG3
*
OUTPUT   OUT       REG3,PA0       *output
```

Figure 9.11   (*continued*)

```
*
*       data record
*
        LAC     RECIC           *recic=recic-1
        SBLK    1
        SACL    RECIC
        BGZ     RTMAIN          *if recic>0 then RTMAIN
*
        LAC     RECNUM          *recic=recnum
        SACL    RECIC
*
        LAC     RECCNT          *reccnt=reccnt-1
        SBLK    1
        BLZ     ENDP            *if reccnt<0 then ENDP
        SACL    RECCNT
*
        LAC     OMEGA
        SACL    *+
        LAC     MODEL
        SACL    *+
        LAC     DISTUR
        SACL    *+
        LAC     REG1
        SACL    *+
        LAC     INTEG
        SACL    *+
        LAC     REG3
        SACL    *+
*       waiting for sampling time
*
RTMAIN  BIOZ    RTMAIN
        B       MAIN
*
*       end of program
*
ENDP    ZAC
        SACL    REG1            *motor 0 volt
        OUT     REG1,PA0
*
        LDPK    32              *DP<--32 page
        LALK    >FFFF
        SACL    ENDMSG          *send PC-9801 end message
*
        RET
```

Figure 9.11 (*continued*)

```
0001                    *
0002                    *           1987/12/
0003                    *                          ASSEMBLER PROGRAM FOR FORCE
                                                   CONTROL
0004                    *
0005                    *
0006                    *
0007                    *
0008        4000    RESULT EQU      >4000          *DATA STORE ADDRESS
0009        0A00    FRCMAX EQU      >0A00          *MAXIMUM VALUE OF
                                                   FORCE
0010        000A    SAMPPT EQU      10             *KE CALCULATING POINT
0011                    *
0012                    *
0013        0000    ERRMSG EQU      0              *ERRMSG ADDRESS
0014        0001    ENDMSG EQU      1              *ENDMSG ADDRESS
0015                    *
0016                    *
0017                    *
0018 0000                      DORG 0
0019                    *
0020                    * CONST
0021                    *
0022 0000          SAMPTM BSS     1                *sampling time
0023 0001          DCOUNT BSS     1                *driving count
0024 0002          FORCED BSS     1                *Fd:target value
0025 0003          PARC   BSS     1                *slope of switching
                   KEXKF  BSS     1                line
0026 0004          KEMIN  BSS     1                *Cstf:stiffness
                   KEMAX  BSS     1                const.
0027 0005          KPGAIN BSS     1                *minimum of ke
0028 0006          DELTA  BSS     1                *maximum of ke
0029 0007          RFERR  BSS     1                *Ks:gain of sl.mode
0030 0008          FFDELT BSS     1                *delta of sl.mode
0031 0009          FVDELT BSS     1                *reference value of
                   OFFSET BSS     1                err
0032 000A                                          *delta of d-force
0033 000B                                          *delta of 1/v
0034 000C                                          *offset of sensor
0035                    *
0036                    * VARIABLES
0037                    *
0038        000D    WRKST  EQU     $
0039                    *
0040 000D          RECPTR BSS     1                *record point of data
                   TIMEFG BSS     1                address
```

Figure 9.12 Force control program

```
0041  000E      MODEL   BSS     1       *f ag of timer
0042  000F      ERR0    BSS     1       *reference model
               ERR1    BSS     1       (step)
0043  0010      CNTRL   BSS     1       *err(k)
0044  0011      REG1    BSS     1       *err(k-1)
0045  0012      REG2    BSS     1       *v(k)
0046  0013      REG3    BSS     1       *register1 for
                                        calculation
0047  0014                              *reg2
0048  0015                              *reg3
0049  0016      FRC0    BSS     1       *f(k)
0050  0017      FRC1    BSS     1       *f(k-1)
0051  0018      VLCTY0  BSS     1       *v(k)
0052  0019      VLCTY1  BSS     1       *v(k-1)
0053  001A      VLCTY2  BSS     1       *v(k-2)
0054  001B      VLCTY3  BSS     1       *v(k-3)
0055  001C      VLCTY4  BSS     1       *v(k-4)
0056  001D      VLCTY5  BSS     1       *v(k-5)
0057  001E      VLCTY6  BSS     1       *v(k-6)
0058  001F      KEAV    BSS     1       *average of ke
0059  0020      KEALL   BSS     1       *sigma(ke(k-j))
0060  0021      KE0     BSS     1       *ke(k)
0061  0022      NUM     BSS     1       *j:number of ke
0062  0023      KFGAIN  BSS     1       *kf:calculated gain
0063  0024      DERR    BSS     1       *err(k)-err(k-1)
0064  0025      SLINE   BSS     1       *switching line
0065  0026      PGAIN   BSS     1       parameter
0066  0027      SSDEL   BSS     1       *kf+ks
0067  0028      SDELTA  BSS     1       *used sl.mode
                                        *used sl.mode

0068            *
0069     0029   QUEST   EQU     $       *start address of que
0070  0029      KEQUE   BSS     40      *que of ke(k)
0071     0050   QUEND   EQU     $-1     *end address of que
0072  0051      DUMMY   BSS     1       *
0073  0052      QNUM    BSS     1       *location of que
0074  0053      QLENG   BSS     1       *length of que
0075  0054      SPNUM   BSS     1
0076            *
0077     0054   WRKEND  EQU     $-1
0078            *
0079            *
0080            *       INIT
0081            *
0082            *
0083  2000              AORG    >2000
```

Figure 9.12 (*continued*)

```
0084 2000 CE01          DINT
0085 2001 CE03          SOVM
0086 2002 CE07          SSXM
0087 2003 CE08          SPM         0
0088 2004 CE04          CNFD
0089                *
0090 2005 C804          LDPK        4
0091 2006 5588          LARP        AR0
0092 2007 D001          LALK        WRKEND
     2008 0054
0093 2009 D003          SBLK        WRKST
     200A 000D
0094 200B 6013          SACL        REG1
0095 200C D001          LALK        WRKST
     200D 000D
0096 200E D002          ADLK        >200
     200F 0200
0097 2010 6014          SACL        REG2
0098 2011 3014          LAR         AR0,REG2
0099 2012 CA00          ZAC
0100 2013 4B13          RPT         REG1
0101 2014 60A0          SACL        *+
0102                *
0103 2015 D001          LALK        RESULT
     2016 4000
0104 2017 600D          SACL        RECPTR
0105                *
0106                *
0107 2018 D001          LALK        QUEND
     2019 0050
0108 201A D003          SBLK        QUEST
     201B 0029
0109 201C 6053          SACL        QLENG
0110 201D D001          LALK        QUEST
     201E 0029
0111 201F D002          ADLK        >200
     2020 0200
0112 2021 6052          SACL        QNUM
0113                *
0114 2022 E012          OUT         CNTRL,PA0
0115                *
0116 2023 D001          LALK        TCHECK
     2024 213D
0117 2025 6013          SACL        REG1
0118 2026 D001          LALK        25
     2027 0019
```

Figure 9.12   (*continued*)

```
0119 2028 5913              TBLW     REG1
0120                 *
0121 2029 2000              LAC      SAMPTM
0122 202A C800              LDPK     0
0123 202B 6002              SACL     2
0124 202C 6003              SACL     3
0125 202D 2004              LAC      4
0126 202E D005              ORK      >0008
     202F 0008
0127 2030 6004              SACL     4
0128                 *
0129 2031 C804              LDPK     4
0130 2032 D001              LALK     >FF60
     2033 FF60
0131 2034 6015              SACL     REG3
0132 2035 E015              OUT      REG3,PA0
0133 2036 8016    ICLOSE IN          FRC0,PA0
0134 2037 2016              LAC      FRC0
0135 2038 100C              SUB      OFFSET
0136 2039 CE23              NEC
0137 203A D003              SBLK     4
     203B 0004
0138 203C F380              BLZ      ICLOSE
     203D 2036
0139 203E CA00              ZAC
0140 203F 6015              SACL     REG3
0141 2040 E015              OUT      REG3,PA0
0142                 *
0143 2041 2002              LAC      FORCED
0144 2042 600F              SACL     MODEL
0145 2043 D001              LALK     SAMPPT
     2044 000A
0146 2045 6054              SACL     SPNUM
0147                 *
0148 2046 CE00              EINT
0149 2047 CE1F              IDLE
0150                 *
0151                 *
0152                 *     MAIN PROGRAM
0153                 *
0154                 *
0155 2048 CE00    MAIN   EINT
0156                 *
0157 2049 8016              IN       FRC0,PA0
0158 204A 2016              LAC      FRC0
0159 204B 100C              SUB      OFFSET
```

Figure 9.12   (continued)

```
0160  204C  CE23            NEC
0161  204D  6016            SACL      FRC0
0162                 *
0163                 *      LAC       DCOUNT
0164                 *      SBLK      >E00        *sampling time = 200H
0165                 *      BNZ       MKMDLA
0166                 *      LAC       FORCED
0167                 *      SACL      MODEL       *MODEL = 800H
0168                 *MKMDLA          LAC       DCOUNT
0169                 *      SBLK      >800        *sampling time = 800H
0170                 *      BNZ       MKMDLB
0171                 *      LALK      >200        *MODEL = 200H
0172                 *      SACL      MODEL
0173                 *
0174                 *make error
0175                 *
0176                 *MKMDLB          LAC       FRC0
0177  204E  100F            SUB       MODEL
0178  204F  6010            SACL      ERR0
0179                 *make derror
0180  2050  1011            SUB       ERR1
0181  2051  6024            SACL      DERR
0182                 *
0183                 *      make adaptive gain by sliding mode
0184                 *
0185                 *sline = parc(6.10)*err0 + derr*2^4.,(12.4)
0186  2052  3C10            LT        ERR0
0187  2053  3803            MPY       PARC
0188  2054  CE14            PAC
0189  2055  CB05            RPTK      5
0190  2056  CE19            SFR
0191  2057  0424            ADD       DERR,4
0192  2058  6025            SACL      SLINE
0193                 *sdelta = ABS(sl) + delta. (12.4)
0194  2059  CE1B            ABS
0195  205A  0008            ADD       DELTA
0196  205B  6028            SACL      SDELTA
0197                 *ssdel = sl/(ABS(sl) + delta). (2.14)
0198  205C  D001            LALK      >4000
      205D  4000
0199  205E  CB0F            RPTK      15
0200  205F  4728            SUBC      SDELTA
0201  2060  D004            ANDK      >7FFF
      2061  7FFF
0202  2062  6013            SACL      REG1
0203  2063  3C13            LT        REG1
```

Figure 9.12   (*continued*)

```
0204 2064 3825          MPY         SLINE
0205 2065 CE14          PAC
0206 2066 6027          SACL        SSDEL
0207                *
0208 2067 2054          LAC         SPNUM
0209 2068 D002          ADLK        1
     2069 0001
0210 206A 6054          SACL        SPNUM
0211 206B D003          SBLK        SAMPPT
     206C 000A
0212 206D F380          BLZ         KESUBJ
     206E 20E3
0213 206F CA00          ZAC
0214 2070 6054          SACL        SPNUM
0215                *
0216                *       make adaptive gain by stiffness
0217                *v(k'-5)...(16.0)     K'=k*10
0218 2071 201D  KESUBA  LAC         VLCTY5
0219 2072 F580          BNZ         KESUBB
     2073 2078
0220 2074 2006          LAC         KEMAX
0221 2075 6021          SACL        KE0
0222 2076 FF80          B           KESUBE
     2077 209E
0223                *
0224 2078 201D  KESUBB  LAC         VLCTY5
0225 2079 CE1B          ABS
0226 207A 6013          SACL        REG1
0227 207B D001          LALK        >4000
     207C 4000
0228 207D CB0F          RPTK        15
0229 207E 4713          SUBC        REG1
0230 207F 000B          ADD         FVDELT
0231 2080 D004          ANDK        >7FFF
     2081 7FFF
0232                *1.0(2.14)/v(k'-5)....(2.14)
0233 2082 6013          SACL        REG1
0234 2083 2016          LAC         FRC0
0235 2084 1017          SUB         FRC1
0236 2085 CE1B          ABS
0237 2086 CB03          RPTK        3
0238 2087 CE18          SFL
0239 2088 000A          ADD         FFDELT
0240 2089 6014          SACL        REG2
0241 208A 3C14          LT          REG2
0242 208B 3813          MPY         REG1
```

Figure 9.12 *(continued)*

```
0243 208C CE14              PAC
0244 208D CE19              SFR
0245 208E CE19              SFR
0246              * ABS(f(k)-f(k-1))*1/v(k'-5)....(0.16)
0247              *    (12.4)*(2.14)=(14.18)
0248 208F 6021              SACL      KE0
0249              *
0250 2090 2021              LAC       KE0
0251 2091 1005              SUB       KEMIN
0252 2092 F480              BGEZ      KESUBD
     2093 2098
0253 2094 2005              LAC       KEMIN
0254 2095 6021              SACL      KE0
0255 2096 FF80              B         KESUBE
     2097 209E
0256              *
0257 2098 2021   KESUBD LAC          KE0
0258 2099 1006              SUB       KEMAX
0259 209A F280              BLEZ      KESUBE
     209B 209E
0260 209C 2006              LAC       KEMAX
0261 209D 6021              SACL      KE0
0262              *
0263 209E 3052   KESUBE LAR          AR0,QNUM
0264 209F 2020              LAC       KEALL
0265 20A0 0021              ADD       KE0
0266 20A1 1080              SUB       *
0267 20A2 CE1B              ABS
0268 20A3 6020              SACL      KEALL
0269 20A4 D003              SBLK      >7FFF
     20A5 7FFF
0270 20A6 F380              BLZ       KESUBF
     20A7 20AB
0271 20A8 D001              LALK      >7FFF
     20A9 7FFF
0272 20AA 6020              SACL      KEALL
0273              *
0274 20AB 2021   KESUBF LAC          KE0
0275 20AC 60A0              SACL      *+
0276              *
0277 20AD 7052              SAR       AR0,QNUM
0278 20AE 2052              LAC       QNUM
0279 20AF D003              SBLK      >200
     20B0 0200
0280 20B1 D003              SBLK      QUEND
     20B2 0050
```

Figure 9.12   *(continued)*

```
0281  20B3  F280            BLEZ      KESUBG
      20B4  20BA
0282  20B5  D001            LALK      QUEST
      20B6  0029
0283  20B7  D002            ADLK      >200
      20B8  0200
0284  20B9  6052            SACL      QNUM
0285                   *
0286  20BA  2022   KESUBG LAC        NUM
0287  20BB  1053            SUB       QLENG
0288  20BC  F180            BGZ       KESUBH
      20BD  20C2
0289  20BE  2022            LAC       NUM
0290  20BF  D002            ADLK      1
      20C0  0001
0291  20C1  6022            SACL      NUM
0292  20C2  2020   KESUBH LAC        KEALL
0293  20C3  CB0F            RPTK      15
0294  20C4  4722            SUBC      NUM
0295              * Keav={Ke(k')+Ke(k'-1)+···+Ke(k'-40)}/40
0296  20C5  601F            SACL      KEAV
0297                   *
0298  20C6  201F   KESUBI LAC        KEAV
0299  20C7  F680            BZ        CONTRL
      20C8  20F3
0300  20C9  D001            LALK      >4000
      20CA  4000
0301  20CB  CB0F            RPTK      15
0302  20CC  471F            SUBC      KEAV
0303              * 1.0(2.14)/Keav(0.16)...(18.-2)
0304  20CD  6013            SACL      REG1
0305                   *
0306                   *    LAC       VLCTY5
0307                   *    SACL      VLCTY6
0308  20CE  201C            LAC       VLCTY4
0309  20CF  601D            SACL      VLCTY5
0310  20D0  201B            LAC       VLCTY3
0311  20D1  601C            SACL      VLCTY4
0312  20D2  201A            LAC       VLCTY2
0313  20D3  601B            SACL      VLCTY3
0314  20D4  2019            LAC       VLCTY1
0315  20D5  601A            SACL      VLCTY2
0316  20D6  2018            LAC       VLCTY0
0317  20D7  6019            SACL      VLCTY1
0318                   *
0319  20D8  2010            LAC       ERR0
```

Figure 9.12 *(continued)*

```
0320  20D9  CE1B              ABS
0321  20DA  1009              SUB         RFERR
0322  20DB  F280              BLEZ        KESUBJ
      20DC  20E3
0323                      *
0324  20DD  3C13              LT          REG1
0325  20DE  3804              MPY         KEXKF
0326  20DF  CE14              PAC
0327  20E0  CB09              RPTK        9
0328  20E1  CE19              SFR
0329  20E2  6023              SACL        KFGAIN
0330                      * Kf=kexkf/Keav....(8.8)
0331                      * (-4.20)*(18.-2)=(16.16)
0332                      *
0333                      *  make pgain
0334  20E3  2010      KESUBJ LAC          ERR0
0335  20E4  F480             BGEZ         GAINMK
      20E5  20EA
0336  20E6  2023             LAC          KFGAIN
0337  20E7  CE23             NEG
0338  20E8  FF80             B            PGMAKE
      20E9  20EB
0339  20EA  2023      GAINMK LAC          KFGAIN
0340  20EB  6026      PGMAKE SACL         PGAIN
0341  20EC  3C27             LT           SSDEL
0342  20ED  3807             MPY          KPGAIN
0343  20EE  CE14             PAC
0344  20EF  CB0D             RPTK         13
0345  20F0  CE19             SFR
0346  20F1  0026             ADD          PGAIN
0347  20F2  6026             SACL         PGAIN
0348                      *
0349                      *make control
0350                      *
0351  20F3  2010      CONTRL LAC          ERR0
0352  20F4  6011             SACL         ERR1
0353  20F5  CE1B             ABS
0354  20F6  6013             SACL         REG1
0355  20F7  3C13             LT           REG1
0356  20F8  3826             MPY          PGAIN
0357  20F9  CE14             PAC
0358  20FA  CB07             RPTK         7
0359  20FB  CE19             SFR
0360  20FC  6012             SACL         CNTRL
0361                      *
```

Figure 9.12 (*continued*)

```
0362  20FD  D003           SBLK      >7FFF
      20FE  7FFF
0363  20FF  F380           BLZ       MIN
      2100  2106
0364  2101  D001           LALK      >7FFF
      2102  7FFF
0365  2103  6012           SACL      CNTRL
0366  2104  FF80           B         OUTPUT
      2105  210D
0367                  *
0368  2106  D102    MIN    ADLK      >7FFF,1
      2107  7FFF
0369  2108  F480           BGEZ      OUTPUT
      2109  210D
0370  210A  D001           LALK      >8000
      210B  8000
0371  210C  6012           SACL      CNTRL
0372                  *
0373                  *
0374  210D  2016  OUTPUT LAC        FRC0
0375  210E  6017           SACL      FRC1
0376  210F  2012           LAC       CNTRL
0377  2110  6018           SACL      VLCTY0
0378                  *
0379  2111  CB03           RPTK      3
0380  2112  CE19           SFR
0381  2113  6015           SACL      REG3
0382  2114  E015           OUT       REG3,PA0
0383                  *
0384                  *
0385  2115  300D           LAR       AR0,RECPTR
0386  2116  2016           LAC       FRC0
0387  2117  60A0           SACL      *+
0388  2118  200F           LAC       MODEL
0389  2119  60A0           SACL      *+
0390  211A  2012                     CNTRL
0391  211B  60A0                     *+
0392              *        LAC       KEAV
0393              *        SACL      *+
0394              *        LAC       QLENG
0395              *        SACL      *+
0396              *        LAC       KE0
0397              *        SACL      *+
0398              *        LAC       DUMMY
0399              *        SACL      *+
0400  211C  2026           LAC       PGAIN
```

Figure 9.12 (*continued*)

```
0401  211D 60A0          SACL        *+
0402  211E 700D          SAR         AR0,RECPTR
0403                *
0404  211F 2016          LAC         FRC0
0405  2120 D003          SBLK        FRCMAX
      2121 0A00
0406  2122 F480          BGEZ        ENDPRG
      2123 212A
0407                *
0408  2124 2001          LAC         DCOUNT
0409  2125 D003          SBLK        1
      2126 0001
0410  2127 6001          SACL        DCOUNT
0411  2128 F580          BNZ         WAIT
      2129 2138
0412                *
0413                *
0414  212A CE01   ENDPRG DINT
0415                *
0416  212B CA00          ZAC
0417  212C 6013          SACL        REG1
0418  212D E013          OUT         REG1,PA0
0419                *
0420  212E C800          LDPK        0
0421  212F 2004          LAC         4
0422  2130 D004          ANDK        >FFF7
      2131 FFF7
0423  2132 6004          SACL        4
0424  2133 C820          LDPK        32
0425  2134 D001          LALK        >FFFF
      2135 FFFF
0426  2136 6001          SACL        ENDMSG
0427                *
0428  2137 CE26          RET
0429                *
0430                *
0431                *
0432  2138 CA00   WAIT   ZAC
0433  2139 600E          SACL        TIMEFG
0434  213A CE1F          IDLE
0435  213B FF80          B           MAIN
      213C 2048
0436                *
0437  213D 200E   TCHECK LAC         TIMEFG
0438  213E F580          BNZ         TMERR
      213F 2144
```

Figure 9.12   *(continued)*

```
0439  2140  D001            LALK      >FFFF
      2141  FFFF
0440  2142  600E            SACL      TIMEFG
0441  2143  CE26            RET
0442               *
0443               *
0444               *
0445  2144  CE01   TMERR    DINT
0446  2145  CA00            ZAC
0447  2146  6013            SACL      REG1
0448  2147  E013            OUT       REG1,PA0
0449  2148  C800            LDPK      0
0450  2149  2004            LAC       4
0451  214A  D004            ANDK      >FFF7
      214B  FFF7
0452  214C  6004            SACL      4
0453  214D  C820            LDPK      32
0454  214E  D001            LALK      >FFFF
      214F  FFFF
0455  2150  6000            SACL      ERRMSG
0456               *
0457  2151  CE1D            POP
0458  2152  CE26            RET
0459               *
0460               *        stop
0461               *
0462  3000                  AORG      >3000
0463               *
0464  3000  CE01            DINT
0465  3001  CE03            SOVM
0466  3002  CE07            SSXM
0467  3003  CE08            SPM       0
0468  3004  CE04            CNFD
0469  3005  C804            LDPK      4
0470  3006  CA00            ZAC
0471  3007  6015            SACL      REG3
0472  3008  E015            OUT       REG3,PA0
0473  3009  CE01            DINT
0474  300A  CE26            RET
0475               *
0476               *        open
0477               *
0478  3010                  AORG      >3010
0479               *
0480  3010  CE01            DINT
0481  3011  CE03            SOVM
```

Figure 9.12 (*continued*)

```
0482  3012  CE07          SSXM
0483  3013  CE08          SPM      0
0484  3014  CE04          CNFD
0485  3015  C804          LDPK     4
0486  3016  D001          LALK     >60
      3017  0060
0487  3018  6015          SACL     REG3
0488  3019  E015          OUT      REG3,PA0
0489  301A  CE01          DINT
0490  301B  CE26          RET
0491              *
0492              *    close
0493              *
0494  3020                AORG     >3020
0495              *
0496  3020  CE01          DINT
0497  3021  CE03          SOVM
0498  3022  CE07          SSXM
0499  3023  CE08          SPM      0
0500  3024  CE04          CNFD
0501  3025  C804          LDPK     4
0502  3026  D001          LALK     >FFA0
      3027  FFA0
0503  3028  6015          SACL     REG3
0504  3029  E015          OUT      REG3,PA0
0505  302A  CE01          DINT
0506  302B  CE26          RET
0507              *
```

Figure 9.12 *(continued)*

For position and velocity sensors, Yasukawa Feedback unit *TFUE-05ZC7* is used.

Speed sensor sensitivity = 7 V/1000 rpm or 3 V/1000 rpm encoder pulse

rate = 500 pulses/revolution

The following brushless servo motor, driver, and tachometers are used for servo motor speed regulation and tracking control:

Shibaura Sheisasusho's brushless servo motor = *ASM-121M*

rated output   = 120 W

rated speed   = 3000 rpm

rated torque   = 3.9 Kgf-cm

power source = PV-1120, servo driver = ADM 121A

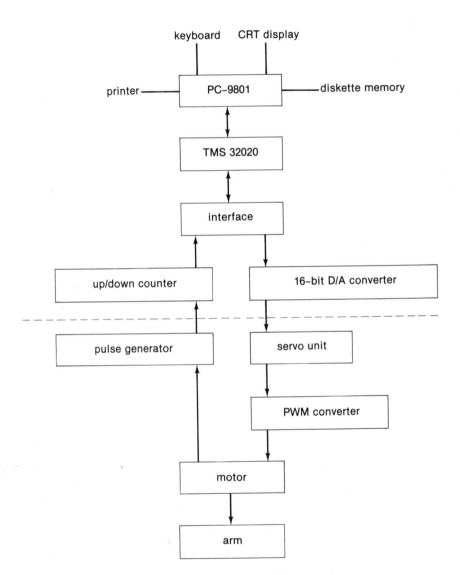

Figure 9.13 Hardware for position control

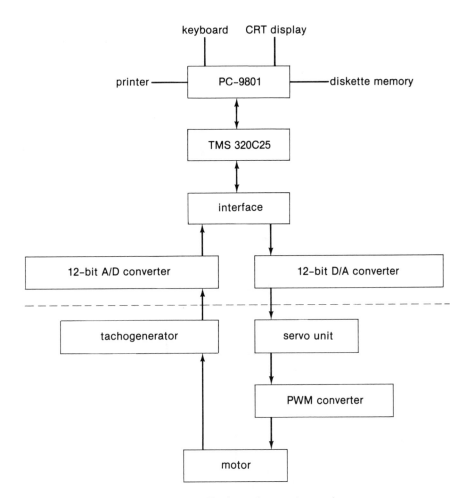

Figure 9.14 Hardware for speed control

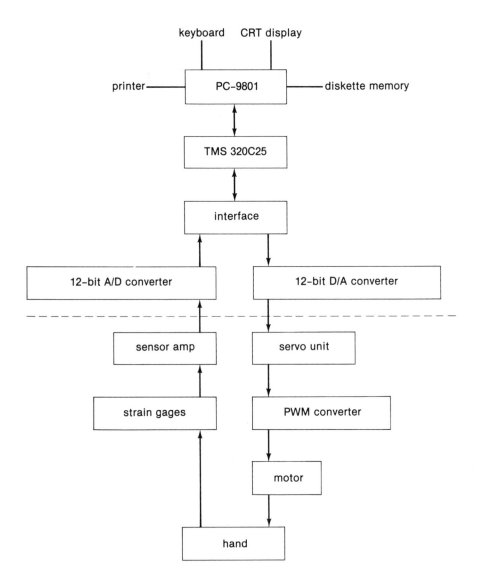

Figure 9.15   Hardware for gripping force control

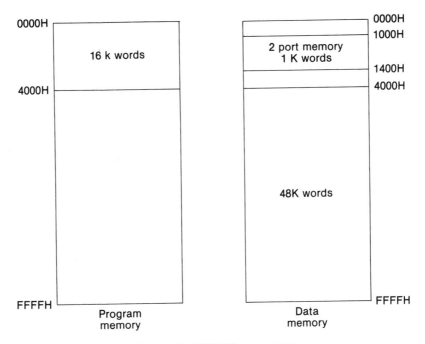

Figure 9.16   TMS 320 memory map

9.4 HARDWARE INTERFACE

Figures 9.17, 9.18, and 9.19 show interface circuits for position, velocity, and force controls, respectively.

Position Control

Feedback pulse counter.   The feedback pulse counter is a 16-bit up/down counter. An encoder generates 500 pulses/revolution. The counter counts the pulses up/down and yields angular position information to the TMS 32020.

D/A converter.   The 16-bit position control is converted to ± 10 V analog voltage, and goes to the servo motor driver.

Timer.   The internal timer of the TMS 320C25 is used instead of an exterior mounted timer.

Speed Control

Scaling amplifier (instrumentation amp.).   Since the input voltage range of the A/D converter is − 5 V through + 5 V, the speed feedback voltage signal

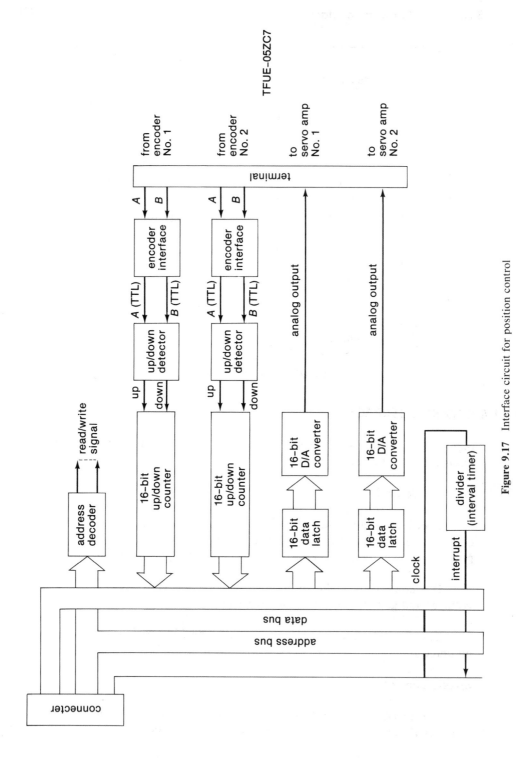

Figure 9.17 Interface circuit for position control

232

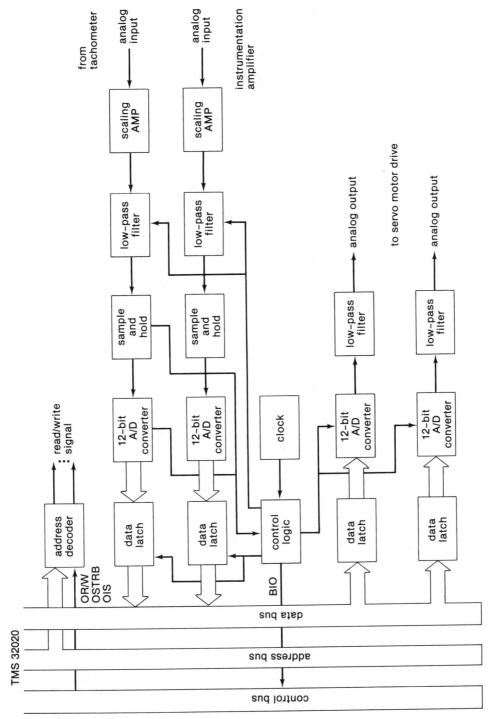

Figure 9.18 Interface circuits for speed control

233

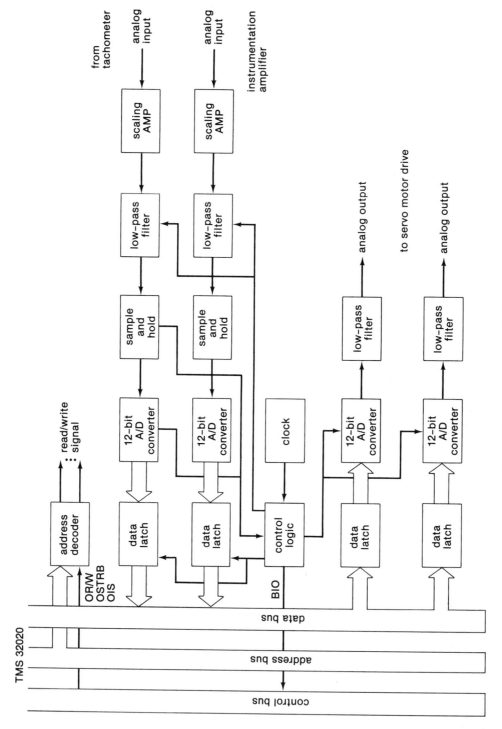

Figure 9.19 Interface circuits for force control

234

from the tachometer is multiplied by $\frac{1}{3}$, so as to be within the input voltage limits of the A/D converter.

Voltage following type low pass filter (instrumentation amp.).   In order to eliminate or reduce noises contained in the signals from the tachometer and the D/A converter, a fourth-order Chebeschef filter is constructed with this instrumentation amplifier.

A/D converter.   The analog feedback signal from the tachometer is converted to the 12-bit digital signal. By the clock signal from the timer, the analog signal is automatically sample-hold and is converted to a digital signal at every clock pulse. After conversion is set to HI, the TMS 32020 is poking BIO and recognizes the conversion completions.

D/A converter.   The 12-bit speed control is converted to ± 5 V analog input voltage to the driver.

Timer.   In order to determine the sampling rate, a timer with a programmable quartz oscillator, SPG8640BN, is constructed. The timer generates clock pulses with frequency ranges from 0.0083 Hz to 1 MHz. The A/D converter starts when the clock pulse comes.

Force Control

Instrumentation amplifier.   This amplifies the small, strain gauge signal. The gain can be adjusted by the dip switch from 4.6 to 400X.

A/D converter.   The analog signal from the strain gauge is converted to 12-bit digital signal with this converter. The clock signal from the timer starts the conversion.

D/A converter.   The 12-bit digital force control is converted to the analog voltage signal (± 5 V) to the driver.

Scaling amplifier.   The force command signal is very small, and in order to obtain high resolution in the D/A converter input voltage, the output of the D/A converter is multiplied by $\frac{1}{4}$ to become 1.25 V.

Timer.   The same timer is used here as the one of 9.4.2.

9.5 SOFTWARE INTERFACE

Position Control

In order to obtain the content of the feedback pulse counter, PA2 bit 1 (ch. 1) or bit 0 (ch. 2) of PA2 is first set, then the counter is latched and held. The latched contents are read from PA0 (ch. 1) or PA1 (ch. 2).

To output the analog voltage, a write is made to PA0 (ch. 1) or PA1 (ch. 2).

Speed Control

A/D conversion.   The output of the A/D is obtained by reading PA0 (ch. 1) or PA1 (ch. 2) while BIO is HI. The A/D converted signal is assigned bits 0 through 11. Bits 12 through 15 take the same content of sign bit 11.

D/A conversion.   To output the analog voltage, a write must be made to the least significant 12-bits of PA0 (ch. 1) or PA1 (ch. 2).

BIO.   BIO is LO during A/D conversion and HI after conversion.

Force Control

A/D, D/A, and BIO are the same as those of section 9.5.2.

9.6 SYSTEM PERFORMANCE

Position Tracking Control

Figure 9.20(a) shows the experimental result for the proposed method. Control is more accurate (robust) in comparison with the result for conventional PI control shown in Fig. 9.20(b).

Speed Regulation and Tracking

Figure 9.21(a) shows the experimental results for the proposed method. Better dynamic response with no overshoot has been obtained, compared with the experimental results for conventional PI control shown in Fig. 9.21(b).

Figure 9.20   (a) Experimental position control result for proposed method.
(b) Experimental position control results for PI control.

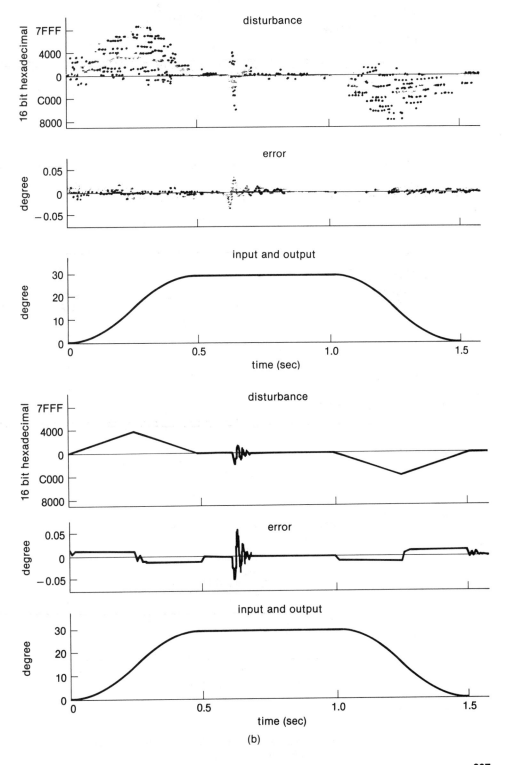

(b)

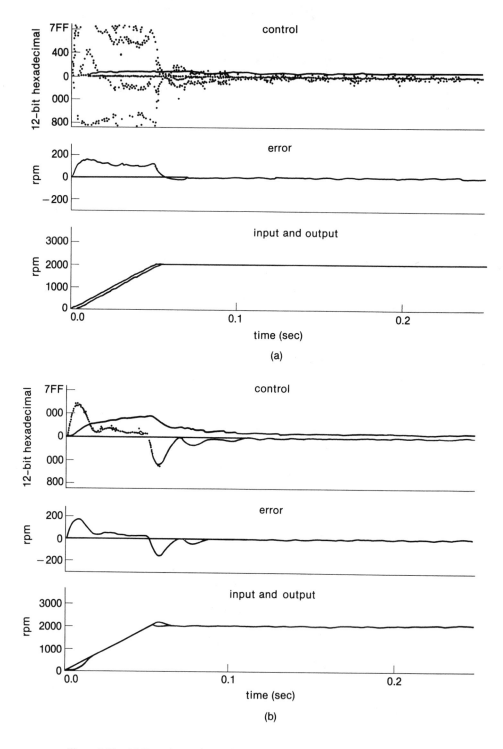

Figure 9.21 (a) Experimental speed control result for proposed method. (b) Experimental speed control result for PI control.

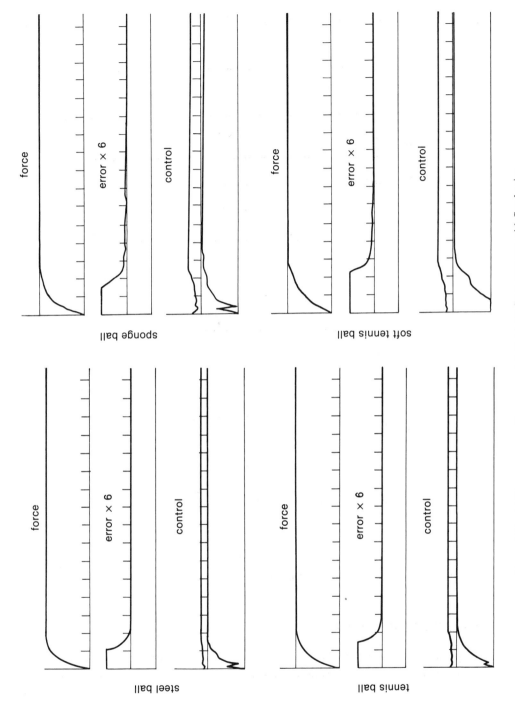

Figure 9.22 (a) Step response with proposed method. (b) Step response with fixed gains.

239

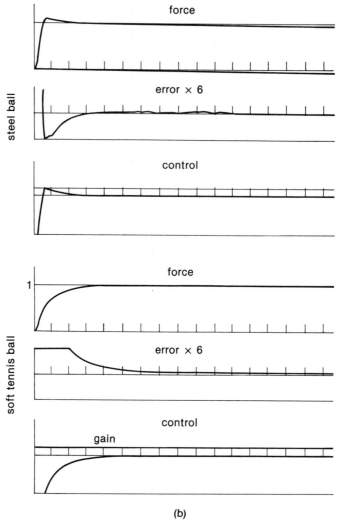

(b)

Figure 9.22 (*continued*)

Gripping Force Control

Experimental results are shown in Figs. 9.22(a) and 9.22(b). The proposed method gives better control performance. Every object is gripped with about the same dynamic response and no overshoot.

Appendix A

TMS 320C25
Instruction Set

Table A–1 lists symbols and abbreviations used in the instruction set summary and in the instruction descriptions. Tables A–2 through A–4 list the complete instruction summary. It is organized according to function. A detailed description of each instruction is listed in the instruction set summary.

TABLE A.1   INSTRUCTION SYMBOLS

Symbol	Meaning
ACC	Accumulator
ARn	Auxiliary Register n (AR0 through AR4 are predefined assembler symbols equal to 0 through 4, respectively.)
ARP	Auxiliary register pointer
B	4-bit field specifying a bit code
CM	2-bit field specifying compare mode
D	Data memory address field
DATn	Label assigned to data memory location n
dma	Data memory address
DP	Data page pointer
FO	Format status bit
I	Addressing mode bit
INTM	Interrupt mode flag bit
K	Immediate operand field
>nn	Indicates nn is a hexadecimal number. (All others are assumed to be decimal values.)
P	Product (P) register
PA	Port address (PA0 through PA15 are predefined assembler symbols equal to 0 through 15, respectively.)
PC	Program counter
PM	2-bit field specifying P register output shift code
pma	Program memory address
PRGn	Label assigned to program memory location n
R	3-bit operated field specifying auxiliary register
S	4-bit left-shift code
T	T register
TOS	Top of stack
X	3-bit accumulator left-shift field
→	Is assigned to
| |	Indicates an absolute value
⟨ ⟩	Items within angle brackets are user-defined
[ ]	Items within brackets are optional
( )	Indicates "contents of"
{ }	Items within braces are alternative items; one of them must be entered.
⟨ ⟩	Angle brackets back-to-back indicate not equal Blanks or spaces must be entered where shown.

TABLE A.2 INSTRUCTION SET SUMMARY

Accumulator Memory Reference Instructions

Mnemonic	Description	No. words	15	14	13	12	11	10	9	8	7	6	5	4	3	2	1	0
ABS	Absolute value of accumulator	1	1	1	0	0	1	1	1	0	0	0	0	1	1	0	1	1
ADD	Add to accumulator with shift	1	0	0	0	0		S			I			D				
ADDH	Add to high accumulator	1	0	1	0	0	1	0	0	0	I			D				
ADDS	Add to low accumulator with sign extension suppressed	1	0	1	0	0	1	0	0	1	I			D				
ADDT†	Add to accumulator with shift specified by T register	1	0	1	0	0	1	0	1	0	I			D				
ADLK†	Add to accumulator long immediate with shift	2	1	1	0	1	S				0	0	0	0	0	0	1	0
AND	AND with accumulator	1	0	1	0	0	1	1	1	0	I			D				
ANDK†	AND immediate with accumulator with shift	2	1	1	0	1	S				0	0	0	0	0	1	0	0
CMPL†	Complement accumulator	1	1	1	0	0	1	1	1	0	0	0	1	0	0	1	1	1
LAC	Load accumulator with shift	1	0	0	1	0		S			I			D				
LACK	Load accumulator immediate short	1	1	1	0	0	1	0	1	0				K				
LACT†	Load accumulator with shift specified by T register	1	0	1	0	0	0	0	1	0	I			D				
LALK†	Load accumulator long immediate with shift	2	1	1	0	1	S				0	0	0	0	0	0	0	1
NEG†	Negate accumulator	1	1	1	0	0	1	1	1	0	0	0	1	0	0	0	1	1
NORM†	Normalize contents of accumulator	1	1	1	0	0	1	1	1	0	1	0	1	0	0	0	1	0
OR	OR with accumulator	1	0	1	0	0	1	1	0	1	I			D				
ORK†	OR immediate with accumulator with shift	2	1	1	0	1	S				0	0	0	0	0	1	0	1
SACH	Store high accumulator with shift	1	0	1	1	0	1		X		I			D				
SACL	Store low accumulator with shift	1	0	1	1	0	0		X		I			D				
SBLK†	Subtract from accumulator long immediate with shift	2	1	1	0	1	S				0	0	0	0	0	0	1	1
SFL†	Shift accumulator left	1	1	1	0	0	1	1	1	0	0	0	0	1	1	0	0	0
SFR†	Shift accumulator right	1	1	1	0	0	1	1	1	0	0	0	0	1	1	0	0	1
SUB	Subtract from accumulator with shift	1	0	0	0	1		S			I			D				
SUBC	Conditional subtract	1	0	1	0	0	0	1	1	1	I			D				
SUBH	Subtract from high accumulator	1	0	1	0	0	0	1	0	0	I			D				
SUBS	Subtract from low accumulator with sign extension suppressed	1	0	1	0	0	0	1	0	1	I			D				
SUBT†	Subtract from accumulator with shift specified by T register	1	0	1	0	0	0	1	1	0	I			D				
XOR	Exclusive-OR with accumulator	1	0	1	0	0	1	1	0	0	I			D				
XORK†	Exclusive-OR immediate with accumulator with shift	2	1	1	0	1	S				0	0	0	0	0	1	1	0
ZAC	Zero accumulator	1	1	1	0	0	1	0	1	0	0	0	0	0	0	0	0	0
ZALH	Zero low accumulator and load high accumulator	1	0	1	0	0	0	0	0	0	I			D				
ZALS	Zero accumulator and load low accumulator with sign extension suppressed	1	0	1	0	0	0	0	0	1	I			D				

TABLE A.2 (*continued*)

Auxiliary Registers and Data Page Pointer Instructions

Mnemonic	Description	No. words	15	14	13	12	11	10	9	8	7	6	5	4	3	2	1	0
CMPR[†]	Compare auxiliary register with auxiliary register AR0	1	1	1	0	0	1	1	1	0	0	1	0	1	0	0	⟨CM⟩	
LAR	Load auxiliary register	1	0	0	1	1	0		R	I				D				
LARK	Load auxiliary register immediate short	1	1	1	0	0	0		R					K				
LARP	Load auxiliary register pointer	1	0	1	0	1	0	1	0	1	1	0	0	0	1			R
LDP	Load data memory page pointer	1	0	1	0	1	0	0	1	0	I			D				
LDPK	Load data memory page pointer immediate	1	1	1	0	0	1	0	0					K				
LRLK1	Load auxiliary register long immediate	2	1	1	0	1	0		R		0	0	0	0	0	0	0	0
MAR	Modify auxiliary register	1	0	1	0	1	0	1	0	1	I			D				
SAR	Store auxiliary register	1	0	1	1	1	0		R	I				D				

[†] These instructions are not included in the TMS32010 instruction set.

TABLE A.3 INSTRUCTION SET SUMMARY (continued)

T Register, P Register, and Multiply Instructions

Mnemonic	Description	No. words	15	14	13	12	11	10	9	8	7	6	5	4	3	2	1	0
APAC	Add P register to accumulator	1	1	1	0	0	1	1	1	0	0	0	0	1	0	1	0	1
LPH†	Load high P register	1	0	1	0	1	0	0	1	1	I			D				
LT	Load T register	1	0	0	1	1	1	1	0	0	I			D				
LTA	Load T register and accumulate previous product	1	0	0	1	1	1	1	0	1	I			D				
LTD	Load T register, accumulate previous product, and move data	1	0	0	1	1	1	1	1	1	I			D				
LTP†	Load T register and store P register in accumulator	1	0	0	1	1	1	1	1	0	I			D				
LTS†	Load T register and subtract previous product	1	0	1	0	1	1	0	1	1	I			D				
MAC†	Multiply and accumulate	2	0	1	0	1	1	1	0	1	I			D				
MACD†	Multiply and accumulate with data move	2	0	1	0	1	1	1	0	0	I			D				
MPY	Multiply (with T register, store product in P register)	1	0	0	1	1	1	0	0	0	I			D				
MPYK	Multiply immediate	1	1	0	1				K									
PAC	Load accumulator with P register	1	1	1	0	0	1	1	1	0	0	0	0	1	0	1	0	0
SPAC	Subtract P register from accumulator	1	1	1	0	0	1	1	1	0	0	0	0	1	0	1	1	0
SPM†	S and P register output shift mode	1	1	1	0	0	1	1	1	0	0	0	0	0	1	0	⟨PM⟩	
SQRA†	Square and accumulate	1	0	0	1	1	1	0	0	1	I			D				
SQRS†	Square and subtract previous product	1	0	1	0	1	1	0	1	0	I			D				

Branch/Call Instructions

Mnemonic	Description	No. words	15	14	13	12	11	10	9	8	7	6	5	4	3	2	1	0
B	Branch unconditionally	2	1	1	1	1	1	1	1	1	I			D				
BACC†	Branch to address specified by accumulator	1	1	1	0	0	1	1	1	0	0	0	1	0	0	1	0	1
BANZ	Branch on auxiliary register not zero	2	1	1	1	1	1	0	1	1	I			D				
BBNZ†	Branch if TC bit ≠ 0	2	1	1	1	1	1	0	0	1	I			D				
BBZ†	Branch if TC bit = 0	2	1	1	1	1	1	0	0	0	I			D				
BGEZ	Branch if accumulator ≥ 0	2	1	1	1	1	0	1	0	0	I			D				
BGZ	Branch if accumulator > 0	2	1	1	1	1	0	0	0	1	I			D				
BIOZ	Branch on I/O status = 0	2	1	1	1	1	1	0	1	0	I			D				
BLEZ	Branch if accumulator ≤ 0	2	1	1	1	1	0	0	1	0	I			D				
BLZ	Branch if accumulator < 0	2	1	1	1	1	0	0	1	1	I			D				
BNV†	Branch if no overflow	2	1	1	1	1	0	1	1	1	I			D				
BNZ	Branch if accumulator ≠ 0	2	1	1	1	1	0	1	0	1	I			D				
BV	Branch on overflow	2	1	1	1	1	0	0	0	0	I			D				
BZ	Branch if accumulator = 0	2	1	1	1	1	0	1	1	0	I			D				
CALA	Call subroutine indirect	1	1	1	0	0	1	1	1	0	0	0	1	0	0	1	0	0
CALL	Call subroutine	2	1	1	1	1	1	1	1	0	I			D				
RET	Return from subroutine	1	1	1	0	0	1	1	1	0	0	0	1	0	0	1	1	0

† These instructions are not included in the TMS32010 instruction set.

Control Instructions

Mnemonic	Description	No. words	15	14	13	12	11	10	9	8	7	6	5	4	3	2	1	0
BIT†	Test bit	1	1	0	0	1			B		I				D			
BITT†	Test bit specified by T register	1	0	1	0	1	0	1	1	1	I				D			
CNFD†	Configure block as data memory	1	1	1	0	0	1	1	1	0	0	0	0	0	0	1	0	0
CNFP†	Configure block as program memory	1	1	1	0	0	1	1	1	0	0	0	0	0	0	1	0	1
DINT	Disable interrupt	1	1	1	0	0	1	1	1	0	0	0	0	0	0	0	0	1
EINT	Enable interrupt	1	1	1	0	0	1	1	1	0	0	0	0	0	0	0	0	0
IDLE†	Idle until interrupt	1	1	1	0	0	1	1	1	0	0	0	1	1	1	1	1	1
LST	Load status register ST0	1	0	1	0	1	0	0	0	0	I				D			
LST1†	Load status register ST1	1	0	1	0	1	0	0	0	1	I				D			
NOP	No operation	1	0	1	0	1	0	1	0	1	0	0	0	0	0	0	0	0
POP	Pop top of stack to low accumulator	1	1	1	0	0	1	1	1	0	0	0	0	1	1	1	0	1
POPD†	Pop top of stack to data memory	1	0	1	1	1	1	0	1	0	I				D			
PSHD†	Push data memory value onto stack	1	0	1	0	1	0	1	0	0	I				D			
PUSH	Push low accumulator onto stack	1	1	1	0	0	1	1	1	0	0	0	0	1	1	1	0	0
ROVM	Reset overflow mode	1	1	1	0	0	1	1	1	0	0	0	0	0	0	0	1	0
RPT†	Repeat instruction as specified by data memory value	1	0	1	0	0	1	0	1	1	I				D			
RPTK†	Repeat instruction as specified by immediate value	1	1	1	0	0	1	0	1	1					K			
RSXM†	Reset sign-extension mode	1	1	1	0	0	1	1	1	0	0	0	0	0	0	1	1	0
SOVM	Set overflow mode	1	1	1	0	0	1	1	1	0	0	0	0	0	0	0	1	1
SST	Store status register ST0	1	0	1	1	1	1	0	0	0	I				D			
SST1†	Store status register ST1	1	0	1	1	1	1	0	0	1	I				D			
SSXM†	Set sign-extension mode	1	1	1	0	0	1	1	1	0	0	0	0	0	0	1	1	1
TRAP†	Software interrupt	1	1	1	0	0	1	1	1	0	0	0	0	1	1	1	1	0

I/O and Data Memory Operations

Mnemonic	Description	No. words	15	14	13	12	11	10	9	8	7	6	5	4	3	2	1	0
BLKD†	Block move from data memory to data memory	2	1	1	1	1	1	1	0	1	I				D			
BLKP†	Block move from program memory to data memory	2	1	1	1	1	1	1	0	0	I				D			
DMOV	Data move in data memory	1	0	1	0	1	0	1	1	0	I				D			
FORT†	Format serial port registers	1	1	1	0	0	1	1	1	0	0	0	0	0	1	1	1	FO
IN	Input data from port	1	1	0	0	0		PA			I				D			
OUT	Output data to port	1	1	1	1	0		PA			I				D			
RTXM†	Reset serial port transmit mode	1	1	1	0	0	1	1	1	0	0	0	1	0	0	0	0	0
RXF†	Reset external flag	1	1	1	0	0	1	1	1	0	0	0	0	0	1	1	0	0
STXM†	Set serial port transmit mode	1	1	1	0	0	1	1	1	0	0	0	1	0	0	0	0	1
SXF†	Set external flag	1	1	1	0	0	1	1	1	0	0	0	0	0	1	1	0	1
TBLR	Table read	1	0	1	0	1	1	0	0	0	I				D			
TBLW	Table write	1	0	1	0	1	1	0	0	1	I				D			

† These instructions are not included in the TMS32010 instruction set.

Appendix B

Digital Control Experimental Procedure

Figure B.1 shows the control tools required for the experiment. In operation, the DSP program written in assembly language is downloaded to the TMS 320C25 from the PC-9801 for execution. The TMS 320C25 has a stored initialization program and data memory areas. Each block of data is transferred from the host computer to the DSP and vice-versa. Experimental data are brought from the DSP to the personal computer for display.

B.1 MAKING SOURCE FILES (PROGRAM WRITING)

A source file is made and modified by using a screen editor. The file name for the screen editor is SED.EXE. Input the following statement:

```
A>SED ⟨rtn⟩
```

Then, your file name is asked for. Input your file name as follows:

```
XXXX.ASM⟨rtn⟩
```

The extension operator .ASM should be added after the file name. The operation instructions will appear on the screen by pressing the "HELP" key. Comments

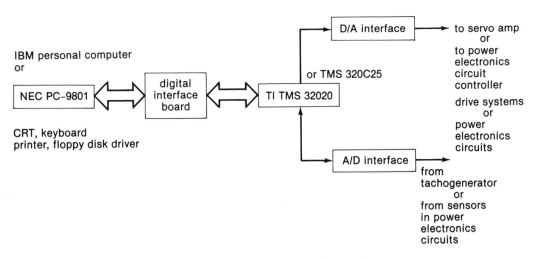

Figure B.1   Control tools for experiment

can be put in after the asterisk (*). You can end the program by writing "^{2}K^{2}X". This closes the file SED. Here, "^{2}" represents pressing the "CTRL" key.

B.2 *ASSEMBLING THE SOURCE FILE*

Your source file can be assembled by writing the cross assembler SAM.EXE as follows:

```
A>XASM ⟨r t n⟩
```

Next, input the file name you are going to assemble.

```
Source File (NUL.ASM)PI = ⟨r t n⟩
Listing File (PI.LST) = ⟨r t n⟩
Object File (PI.MPO) = ⟨r t n⟩
```

Note that the source file name becomes the default name for the listing and object files; the source can be used (entered) by just pressing the return key.

B.3 *DOWNLOADING PROGRAM TO THE TMS 320 FROM THE HOST COMPUTER AND DISPLAY OF EXPERIMENTAL RESULTS*

First, turn on the power, then insert a floppy disk. The following statement appears on the display:

```
Microsoft MS-DOS Version 3.10
Copyright 1981.85 Microsoft Corp./NEC corporation
Command Version 3.10
Today is 1987-07-10 (Friday)
Input today's date:
```

Now, press the return key two times. Then, you wait for the command.

```
Microsoft MS-DOS version 3.10
Copyright 1981.85 Microsoft Corp./NEC corporation
Command Version 3.10
Today is 1987-07-10 (Friday)
Input today's date:
It is 15:33:00.00 now
Input present time:
A>
```

You write the following load N88 BASIC.

```
A>N99BASIC ⟨rtn⟩
NEC N-88 BASIC (86) Version 1.0
Copyright(c) 1984 by NEC Corporation
220324 Bytes free
OK
```

Here you run the program as follows:

```
RUN ''SRVIPL.BAS'' ⟨rtn⟩
```

The program SRVIPL.BAS is loaded and executed by this command. The program is downloaded to the TMS 32020. After loading, the "SRVAID.BAS" (experimental results display program) is executed automatically.

Appendix C

Programs for Handling Experimental Data

The following four programs are used to download the control program to the TMS 320C25 and to handle the experimental data:

SRVIPL.BAS	MS-DOS N88 BASIC
SRVAID.BAS	MS-DOS N88 BASIC
SRVAID.MPO	TMS320C25 ASSEMBLER
SRVAID.MAC	8086 ASSEMBLER

The four programs are explained here.

SRVIPL.BAS

This is the program with which SRVAID.MAC, SRVAID.MPO, and control program PI.MPO are downloaded, or loaded. After loading MS-DOS N88 BASIC, this program is executed; then these programs are loaded. SRVAID.BAS is continuously, automatically executed.

SRVAID.BAS

This program is used for declaring control data, downloading, and executing the control program and obtaining and displaying experimental data.

SRVAID.MPO

This program is for the purpose of initializing the TMS 320C25, interrupt service, and transferring data to the TMS 320C25.

SRVAID.MAC

This is the subroutine with which the work area of the host computer is cleared and data is transferred between the 2-port memory on the TMS 320 data memory and the work area on the host computer memory. This is written in machine language for the 9801 assembler.

SRVAID.MAC consists of three subroutines: MEMCLR, TWSEND, and WTSEND.

MEMCLR is the program for clearing (writing zeros) the work area whose addresses are 000H through FFFH on the segment 5000H. You just write:

CALL MEMCLR

TWSEND is the subroutine to transfer data from the 2-port memory on the TMS 320 to the work area of the host computer. You can use the form:

CALL TWSEND (PCSOAD PCDIAD)

PSCOAD is the integer variable representing the origin address # of the 2-port memory (segment C000H). PCDIAD is the integer variable representing the destination address # of the work area (segment 5000H).
WTSEND is the subroutine which transfers data from the work area to the 2-port memory. You call the subroutine as:

CALL WTSEND (ADRS, PCDIAD)

ADRS represents the address # of the work area. PCDIAD shows the address # on the 2-port memory.

SRVIPL.BAS

The program is listed in Fig. C.1.

Statement #	Function
70$240	Variable definition MSUB machine language subroutine WAREA work area segment TWOPORT 2-port memory segment
110$130	Machine language subroutine address determination MEMCLR work area clearance WTSEND transfer from work area to 2-port memory

```
10  '             SAVE "SRVIPL.BAS"
20  '             1987.4.7
30  '
40   CLEAR &H1010
50   DEFINT A-Z
60  '
70   MSUB   = &H4FF0
80   WAREA  = &H5000
90   TWOPORT= &HC000
100 '
110   MEMCLR= &H0
120   WTSEND= &H20
130   TWSEND= &H60
140 '
150   SDMADRL= &H7F8     : SDMADRH= &H7F9
160   DDMADRL= &H7FA     : DDMADRH= &H7FB
170   RPTCRGL= &H7FC     : RPTCRGH= &H7FD
180   PCTOTML= &H7FE     : PCTOTMH= &H7FF
190 '
200   DSTODSL= &H50      : DSTODSH= &H0
210   DSTOPSL= &H61      : DSTOPSH= &H0
220   CTLADRS= &H800     : CTLDATA= 0
230 '
240   NAM$= "PI     .MPO"    : 'control program name
250   GOSUB *IPLSUB         : 'load and run initial program
260   GOSUB *CNTPGLOAD      : 'load control program
270   PRINT "chain 'SRVAID.BAS'"
280   CHAIN "SRVAID.BAS"
290 '
300 '************************
310 '* 1'st step subroutines *
320 '************************
330 '
340 *IPLSUB
350 'Initial Program loading
360   CLS 3
370   PRINT "initial program loading"
380   PRINT "loading SRVAID.MAC"
390   DEF SEG=MSUB
400   BLOAD "SRVAID.MAC"
410   DEF SEG=WAREA
420   PRINT "loading SRVAID.MPO"
430   GOSUB *RESETSUB
440   GOSUB *IPLMAPSUB
450   DEF SEG=TWOPORT
460   K=14
```

Figure C.1   Program SRVIPL.BAS

```
470    OPEN "SRVAID.MPO" FOR INPUT AS #1
480    INPUT #1,A$
490    B$=MID$(A$,K,5)
500    C$=LEFT$(B$,1)
510    D$=RIGHT$(B$,4)
520    IF C$=":" GOTO 660
530    IF C$="B" GOTO 570
540    IF C$="9" THEN ADRS=VAL("&H"+D$):PX=0:GOTO 620
550    IF C$<>"7" THEN PRINT "error 2":END
560    GOTO 640
570    LOW  =VAL("&H"+RIGHT$(D$,2))
580    HIGH =VAL("&H"+LEFT$(D$,2))
590    POKE ADRS*2,LOW
600    POKE ADRS*2+1,HIGH
610    ADRS=ADRS+1
620    K=K+5
630    GOTO 490
640    K=1
650    IF EOF(1)=0 GOTO 480
660    CLOSE #1
670    GOSUB *STARTSUB        :FOR I=1 TO 100:NEXT I
680    GOSUB *RESETSUB
690    GOSUB *RUNMAPSUB       :FOR I=1 TO 100:NEXT I
700    GOSUB *STARTSUB        :FOR I=1 TO 100:NEXT I
710    DEF SEG=WAREA
720    RETURN
730    '
740    *CNTPGLOAD
750    'control program loading
760    PRINT "control program loading"
770    PRINT "loading ";NAM$
780    DEF SEG=MSUB
790    CALL MEMCLR            :'clear work area
800    DEF SEG=WAREA
810    PCSOAD=&H7FFF
820    PCENAD=&H0
830    K=14
840    OPEN NAM$ FOR INPUT AS #1
850    INPUT #1,A$
860    B$=MID$(A$,K,5)
870    C$=LEFT$(B$,1)
880    D$=RIGHT$(B$,4)
890    IF C$=":" GOTO 1050
900    IF C$="B" GOTO 940
910    IF C$="9" THEN ADRS=VAL("&H"+D$):GOTO 1010
920    IF C$<>"7" THEN PRINT "error 2":END
```

Figure C.1 *(continued)*

```
930    GOTO 1030
940    IF PCSOAD>ADRS*2 THEN PCSOAD=ADRS*2      :'PCSOAD =source
address on PC
950    IF PCEDAD<ADRS*2 THEN PCEDAD=ADRS*2+1  :'PCEDAD
=destination address on PC
960    LOW  =VAL("&H"+RIGHT$(D$,2))
970    HIGH =VAL("&H"+LEFT$(D$,2))
980    POKE ADRS*2,LOW
990    POKE ADRS*2+1,HIGH
1000   ADRS=ADRS+1
1010   K=K+5
1020   GOTO 860
1030   K=1
1040   IF EOF(1)=0 GOTO 850
1050   CLOSE #1
1060   PRINT
1070   TMDIAD=PCSOAD/2
1080   GOSUB *PCTOPSSUB
1090   RETURN
1100   '
1110   '***********************
1120   '* 2'nd step subroutines  *
1130   '***********************
1140   '
1150   *STARTSUB
1160   '
1170   CTLDATA=CTLDATA OR &H80
1180   DEF SEG=TWOPORT
1190   POKE CTLADRS,CTLDATA
1200   DEF SEG=WAREA
1210   RETURN
1220   '
1230   *RESETSUB
1240   '
1250   CTLDATA=CTLDATA AND &H7F
1260   DEF SEG=TWOPORT
1270   POKE CTLADRS,CTLDATA
1280   DEF SEG=WAREA
1290   RETURN
1300   '
1310   *RUNMAPSUB
1320   '
1330   CTLDATA=CTLDATA OR &H40
1340   DEF SEG=TWOPORT
1350   POKE CTLADRS, CTLDATA
1360   DEF SEG=WAREA
```

Figure C.1 (*continued*)

```
1370   RETURN
1380   '
1390   *IPLMAPSUB
1400   '
1410   CTLDATA=CTLDATA AND &HBF
1420   DEF SEG=TWOPORT
1430   POKE CTLADRS,CTLDATA
1440   DEF SEG=WAREA
1450   RETURN
1460   '
1470   *PCTOPSSUB
1480   '
1490   PCDIAD=&H0
1500   BLEN=&HFF
1510   FOR ADRS=PCSOAD TO PCEDAD STEP &H200
1520      ADRS$=RIGHT$("000"+HEX$(TMDIAD),4)
1530      ADRSL=VAL("&H"+RIGHT$(ADRS$,2))
1540      ADRSH=VAL("&H"+LEFT$(ADRS$,2))
1550      DEF SEG=MSUB
1560      CALL WTSEND(ADRS,PCDIAD)
1570      DEF SEG=TWOPORT
1580      POKE SDMADRL,0
1590      POKE SDMADRH,&H10
1600      POKE DDMADRL,ADRSL
1610      POKE DDMADRH,ADRSH
1620      POKE RPTCRGL,BLEN
1630      POKE RPTCRGH,0
1640      POKE PCTOTML,DSTOPSL
1650      POKE PCTOTMH,DSTOPSH
1660      TMDIAD=TMDIAD+&H100
1670   NEXT ADRS
1680   DEF SEG=WAREA
1690   RETURN
1700   '
1710   *PCTODSSUB
1720   '
1730   PCDIAD=&H0
1740   BLEN=&HFF
1750   FOR ADRS=PCSOAD TO PCEDAD STEP &H200
1760      ADRS$=RIGHT$("000"+HEX$(TMDIAD),4)
1770      ADRSL=VAL("&H"+RIGHT$(ADRS$,2))
1780      ADRSH=VAL("&H"+LEFT$(ADRS$,2))
1790      DEF SEG=MSUB
1800      CALL WTSEND(ADRS,PCDIAD)
1810      DEF SEG=TWOPORT
1820      POKE SDMADRL,0
```

Figure C.1   (*continued*)

```
1830      POKE SDMADRH,&H10
1840      POKE DDMADRL,ADRSL
1850      POKE DDMADRH,ADRSH
1860      POKE RPTCRGL,BLEN
1870      POKE RPTCRGH,0
1880      POKE PCTOTML,DSTODSL
1890      POKE PCTOTMH,DSTODSH
1900      TMDIAD=TMDIAD+&H100
1910   NEXT ADRS
1920   DEF SEG=WAREA
1930   RETURN
```

Figure C.1 (*continued*)

SRVIPL.BAS

The program is listed in Fig. C.1.

Statement #	Function
70–240	Variable definition MSUB machine language subroutine WAREA work area segment TWOPORT 2-port memory segment
110–130	Machine language subroutine address determination MEMCLR work area clearance WTSEND transfer from work area to 2-port memory TWSEND transfer from 2-port memory to work area
150–180	Register address determination SDMADRL,H Source address register DDMADRL,H Destination address register RPTCRGL,H Repeating counter register PCTOTML,H Command register
200–210	Command definition DSTODSL,H Command transferring data from DS to DS DSTOPSL,H Command transferring data from DS to DS
220	Control data and control register determination
240	Control program file name definition
250	Loading initial program, downloading to the TMS 320 and execution
260	Loading the control program and downloading it to the TMS 320C25
280	Chain "SRMAID.BAS" IPLSUB
340–720	Initial program loading and downloading to the TMS 320C25; execution
390–400	Loading SRVAID.MAC, the machine language subroutine
430–440	TMS 320C25 resetting and memory remapping for the IPL mode

Statement #	Function
460–660	Loading SRVAID.MPO * There is a detailed explanation later in the TMS 320C25 assembler object file section.
670	SRVAID.MPO execution MPO is transferred to large S-RAM (8K words).
680–700	TMS 320C25 resetting, and memory remapping for the RUN mode. SRVAID.MPO is now located from 000H of the DS. After execution the TMS 320 waits for the interrupt.
740–1090	CNTPGLOAD Loading the control program, then downloading to PS of the TMS 320.
830–1040	Loading the control program. Statements # 850, 860, 990, 1000 are for checking initial and final addresses of the control program.
1070–1080	PCSOAD contains the program's initial address. PCEDAD contains program's final address. Control program is transferred to PS of TMS 320C25 by loading the destination address of PS in the TMS 350C25 and calling PC to PSSUB.
1150–1210	STARTSUB Setting bit 7 of the control register sends the TMS 320C25 into the RUN mode.
1230–1290	RESETSUB Resetting bit 7 of the control register sends the TMS 320C25 into the reset mode.
1310–1370	RUN MAPSUB Setting bit 6 of the control register configures the memory map for the RUN mode.
1390–1450	IPLMAPSUB Resetting bit 6 of the control register configures the memory map for the IPL mode.
1470–1690	PCTOPSSUB Data of the address in PCSOAD through address in PCEDAD in work area of host computer are transferred to the address in TMDIAD of the TMS 320C25's PS.
1550–1560	256 words of the host computer's work area are transferred to 2-port memory.
1580–1650	All registers are set in the TMS 320C25 and they exchange data between each other.
1710–1930	PCTODSSUB Data in address PCSOAD through PCEPAD in host computer work area are transferred to address TMDIAD in the DS of TMS 320C25.
1790–1800	256 words of host computer work area are transferred to 2-port memory.
1820–1890	All registers are set in the TMS 320C25 and they exchange data between each other.

SRVAID.BAS

```
10  '              SAVE "SRVAID.BAS"
20  '              1987.2.1
30  '              aid tool for using DSP TMS 32020
40  '
50  '***************
60  '* MAIN ROUTINE *
70  '***************
80  CLEAR &H1010          :'value for 384Kbyts memory
90  CLS
100  PRINT "Running now  !"
110  GOSUB *INITTMS        :'program and TMS initialization
120  GOSUB *CNTDATSND      :'constnt's sent to TMS
130  GOSUB *CNTPGRUN       :'run control program
140  GOSUB *RSLTGET        :'get results
150  GOSUB *MAKGRPH        :'display obtained results
160  A$=INKEY$
170  IF A$="" GOTO 160
180  CONSOLE 0,25,0,1
190  END
200  END
210  '
220  '*********************
230  '* 1'st step subroutine *
240  '*********************
250  '
260  ' ---------------------------------------------------------
270  *INITTMS
280  '
290  MSUB    =&H4FF0
300  WAREA   =&H5000
310  TWOPORT =&HC000
320  '
330  MEMCLR=&H0
340  WTSEND=&H20
350  TWSEND=&H60
360  '
370  RUNADRL=&H7F6     :RUNADRH=&H7F7
380  SDMADRL=&H7F8     :SDMADRH=&H7F9
390  DDMADRL=&H7FA     :DDMADRH=&H7FB
400  RPTCRGL=&H7FC     :RPTCRGH=&H7FD
410  PCTOTML=&H7FE     :PCTOTMH=&H7FF
420  '
430  DSTODSL=&H50      :DSTODSH=&H0
440  RUNINGL=&H68      :RUNINGH=&H0
```

Figure C.2

```
450   RETURN
460 ' ----------------------------------------------------
470   *CNTDATSND
480 '    control data sent to TMS
490 '    clear work area
500   DEF SEG=MSUB
510   CALL MEMCLR
520   DEF SEG=WAREA
530 '    set constants for control
540   I=&HA000
550   STD=VAL("&H0020"):GOSUB *DS:'         sampling time
560   STIME=STD
570   STD=VAL("&H1000"):GOSUB *DS:'         drive count
580   STD=VAL("&H4000"):GOSUB *DS:'         reference value
590   STD=VAL("&H0010"):GOSUB *DS:'         P gain
600   STD=VAL("&H0200"):GOSUB *DS:'         I gain
610   STD=VAL("&H0000"):GOSUB *DS:'         offset
620 '
630 '    constant's for control sent to TMS
640   PCSOAD=&HA000      :'constants initial address (on PC
address)
650   PCEDAD=&HA018      :'constants end address (on PC address)
660   TMDIAD=&H200       :'destination address (on TMS address)
670   GOSUB *PCTODSSUB
680   RETURN
690   *DS
700   IF STD<0 THEN STD=STD+65536!
710   A=INT(STD/256):B=STD-A*256
720   POKE I+1,A:POKE I,B
730   I=I+2
740   RETURN
750 ' ----------------------------------------------------
760   *CNTPGRUN
770 '    run control program on TMS
780   RUNADS$="2000"     :'control program running address
790   DEF SEG=TWOPORT
800   POKE 0,0:POKE 1,0:POKE 2,0:POKE 3,0
810   POKE RUNADRL,VAL("&H"+RIGHT$(RUNADS$,2))
820   POKE RUNADRH,VAL("&H"+LEFT$(RUNADS$,2))
830   POKE PCTOTML,RUNINGL
840   POKE PCTOTMH,RUNINGH
850 '    wait for job end and clear 2-port memory
860   ERRMSG=PEEK(0)
870   ENDMSG=PEEK(2)
880   IF ERRMSG=255 THEN PRINT "Error    Sampling time is too short
!'":END
```

Figure C.2 *(continued)*

```
890   IF ENDMSG=0 THEN GOTO 860
900   DEF SEG=WAREA
910   RETURN
920 ' ----------------------------------------------------------
930 *RSLTGET
940 '    result data get from DSP data memory
950   DEF SEG=WAREA
960   TMSDAD=&H4000        :'start of data area (on TMS address)
970   TMEDAD=&H8000        :'end of data area (on TMS address)
980   PCDIAD=&H0           :'destination address (on PC address)
990 '
1000   GOSUB *DSTOPCSUB
1010   RETURN
1020 ' ----------------------------------------------------------
1030 *MAKGRPH
1040   DEF SEG=WAREA
1050   SCREEN 3,0,0,1
1060   CONSOLE 0,23,0,1
1070   WINDOW (-300,-6016)-(4000,400)
1080   CLS 3
1090   YERR=-2816 : YCNTR=-4864
1100   LINE(0,     0)-)4000,-1920),1,B
1110   LINE(0,-2048)-(4000,-3584),1,B
1120   LINE(0,-3712)-(4000,-6016),1,B
1130   LINE(0, YERR)-(4000, YERR),5
1140   LINE(0,YCNTR)-(4000,YCNTR),5
1150 '
1160   TML=50000!/(STIME*.8)
1170   FOR X=0 TO 4000 STEP TML              .05 SEC
1180     LINE(X-1,     0)-(X+1,      50),7,BF
1190     LINE(X-1, YERR)-(X+1, YERR+50),7,BF
1200     LINE(X-1,YCNTR)-(X+1,YCNTR+50),7,BF
1210   NEXT X
1220 '
1230   FOR Y=0 TO 1536 STEP 512           :' 512=1000rpm
1240     LINE(-20,-Y+8)-(0,-Y-8),7,BF
1250   NEXT Y
1260   FOR Y=2304 TO 6016 STEP 512
1270     LINE(-20,-Y+8)-(0,-Y-8),7,BF
1280   NEXT Y
1290 '
1300   LOCATE 0,23 : PRINT"     0"
1310   LOCATE 0,0
1320   PRINT" 7FFF"
1330   PRINT"     "
1340   PRINT" 4000"
```

Figure C.2 *(continued)*

```
1350   PRINT"        "
1360   PRINT"     0"
1370   PRINT"        "
1380   PRINT" C000"
1390   PRINT"        "
1400   PRINT" 8000"
1410   PRINT"        "
1420   PRINT"  200"
1430   PRINT"        "
1440   PRINT"     0"
1450   PRINT"        "
1460   PRINT" -200"
1470   PRINT"        "
1480   PRINT"        "
1490   PRINT" 3000"
1500   PRINT"        "
1510   PRINT" 2000"
1520   PRINT"        "
1530   PRINT" 1000"
1540  '
1550     FOR SAMPL=0 TO 4000 STEP 10
1560       I=SAMPL*8
1570  '
1580       VELO=VAL("&H"+HEX$(PEEK(1+1)*256+PEEK(1)))/16
1590       MPOS=VAL("&H"+HEX$(PEEK(1+3)*256+PEEK(I+2)))/16
1600       PCNT=VAL("&H"+HEX$(PEEK(I+5)*256+PEEK(I+4)))/16
1610       ICNT=VAL("&H"+HEX$(PEEK(I+7)*256+PEEK(I+6)))/16
1620       PERR=MPOS-VELO
1630  '
1640       IF ABS(PCNT)<2300 THEN PSET(SAMPL,-PCNT/2+YCNTR),4
1650       IF ABS(ICNT)<2300 THEN PSET(SAMPL,-ICNT/2+YCNTR),6
1660       IF ABS(PERR)<153   THEN PSET(SAMPL,-PERR*5+YERR,3
1670       PSET(SAMPL,-MPOS),3
1680       PSET(SAMPL,-VELO),6
1690  '
1700     NEXT SAMPL
1710     RETURN
1720  '
1730  '    *********************
1740  '    * 2'nd step subroutine *
1750  '    *********************
1760  '
1770  *PCTODSSUB
1780   IF PCSOAD<0 THEN PCSOAD=PCSOAD+65536!
1790   IF PCEDAD<0 THEN PCEDAD=PCEDAD+65536!
1800   IF TMDIAD<0 THEN TMDIAD=TMDIAD+65536!
```

Figure C.2 (*continued*)

```
1810  '
1820   PCDIAD%=&H0
1830   BLEN=&HFF
1840   FOR ADRS=PCSOAD TO PCEDAD STEP &H200
1850     ADRS$=RIGHT$("000"+HEX$(TMDIAD),4)
1860     ADRSL=VAL("&H"+RIGHT$(ADRS$,2))
1870     ADRSH=VAL("&H"+LEFT$(ADRS$,2))
1880     DEF SEG=MSUB
1890     ADRS%=VAL("&H"+HEX$(ADRS))
1900     CALL WTSEND(ADRS%,PCDIAD%)
1910     DEF SEG=TWOPORT
1920     POKE SDMADRL,0
1930     POKE SDMADRH,&H10
1940     POKE DDMADRL,ADRSL
1950     POKE DDMADRH,ADRSH
1960     POKE RPTCRGL,BLEN
1970     POKE RPTCRGH,0
1980     POKE PCTOTML,DSTODSL
1990     POKE PCTOTMH,DSTODSH
2000     TMDIAD=TMDIAD+&H100
2010   NEXT ADRS
2020   DEF SEG=WAREA
2030   RETURN
2040  ' ------------------------------------------------------
2050  *DSTOPCSUB
2060  '
2070  IF TMSOAD<0 THEN TMSOAD=TMSOAD+65536!
2080  IF TMEDAD<0 THEN TMEDAD=TMEDAD+65536!
2090  IF PCDIAD<0 THEN PCDIAD=PCDIAD+65536!
2100   PCSOAD%=&H0
2110   BLEN=&HFF
2120   FOR ADRS=TMSOAD TO TMEDAD STEP &H100
2130     ADRS$=RIGHT$("000"+HEX$(ADRS),4)
2140     ADRSL=VAL("&H"+RIGHT$(ADRS$,2))
2150     ADRSH=VAL("&H"+LEFT$(ADRS$,2))
2160     DEF SEG=TWOPORT
2170     POKE SDMADRL,ADRSL
2180     POKE SDMADRH,ADRSH
2190     POKE DDMADRL,0
2200     POKE DDMADRH,&H10
2210     POKE RPTCRGL,BLEN
2220     POKE RPTCRGH,0
2230     POKE PCTOTML,DSTODSL
2240     POKE PCTOTMH,DSTODSH
2250     DEF SEG=MSUB
```

Figure C.2 (*continued*)

```
2260    PCDIAD%=VAL("&H"+HEX$(PCDIAD))
2270    CALL TWSEND(PCSOAD%,PCDIAD%)
2280    PCDIAD=PCDIAD+&H200
2290  NEXT ADRS
2300  DEF SEG=WAREA
2310  RETURN
```

Figure C.2   (*continued*)

SRVAID.BAS

The program is listed in Fig. C.2.

Statement #	Function
100	Variable definition
110	Transferring control data to TMS 320C25
120	Control program execution
130	Obtaining experimental data
140	Displaying experimental data
270–450	Variable definition MSUB Machine language subroutine segment WAREA work area segment TWOPORT 2-port memory segment
330–350	Machine language subroutine address determination MEMCLR Work area clearance WTSEND Transfer from work area to 2-port memory TWSEND Transfer from 2-port memory to work area
370–410	Register address determination RUNADRL,H Execution start address register SDMADRL,H Source address register DDMADRL,H Destination address register PCTOTML,H Command register
430–440	Command determination DSTODSL,H Transfer of command from DS to DS RUNINGL,H Program execution start command
470–740	Control data transfer to TMS 320C25
500–510	Work area clearance
540–610	Control data
640–670	Downloading control data to TMS 320C25
690–740	Control data set into memory
760–910	Control program execution
780	Start of execution; address determination
830–840	Start of execution; address set in command register and control program execution

Statement #	Function
860–890	End of execution detection With ENDMSG = 255, Normal execution end With ERRMSG = 255, Error occurred, execution stop
930–1010	Data acquisition Here data obtained are transferred to work area by setting initial address in TMSSOAD, last address on TMEDAD and destination address on work area of host computer and by calling DSTOPCSUB.
1030–1710	Experimental result display
1050–1080	Screen mode set and screen clearance
1090 = 1530	Coordinate axes and screen clear
1550–1700	Graphic display VELO: Speed (yellow) MPOS: Command or reference (purple) PCNT: Proportional control (green) ICNT: Integral control (yellow) PERR: Error (purple)
1770–2030	PCTODSSUB Data on address PCSOAD through PCEDAD of work area are transferred to address TMDIAD of PS in the TMS 320C25.
1880–1900	256 words on work area of host computer are transferred to 2-port memory.
1910–1990	All registers set. Data transfer inside the TMS 320C25
2050–2310	DSTOPCSUB Data on address TMSOAD through TMEDAD of the TMS 320C25 DS are transferred to the address in PCDIAD.
2160–2240	Data transfer from 2-port memory to work area

SRVAID.MPO

The program is listed in Fig. C.3.

0001	IDT set
0005–0012	Label definition
0033–0036	Data at address OH through OFFH of program memory is transferred to address 8000H through 80FFH of data memory.
00337–0039	INTO mask removal
0041–0043	Interrupt permission and waiting for interrupt
0049–0052	INTO interrupt processing Branching to address contained in data memory as DCTOTU
0054–0059	Interrupt processing for INTI through TRAP Return without any execution
0065–0093	Command processing routine
0065–0072	DSTODS Transfer routine between data memories of the TMS 320C25; Operand of BLKD (0071) is replaced by contents of SDMADR. ARO set at destination address (0069), and repeat counter set at contents of RPTCRG (0070), transfer (0071)

Statement #	Function
0074–0080	PSTODS
	Routine data transfer from program memory to data memory
0074–0076	ACC set at source address;
0077	ARO set at destination address;
0078	Repeat counter set at contents of RPTCRG;
0079	Data transfer
0082–0088	DSTOPS
	routine for data transfer data memory to program memory
0082–0084	ACC set at destination address;
0085	ARO set at source address;
0086	and repeat counter is set at contents of RPTCTG;
0087	Data transfer
0090–0095	RUNPRG
	Control program execution routine
	Branch to the address contained in RUNADR

SSRVAID.MAC

The program is listed in Fig. C.4.

0000–0016	*MEMCLR
0000–0003	Register evacuation
0004–0008	setting ES at 5000H
0009–000C	clear AX,DI
000D–0010	setting # of repeats at 8000H
0011	ES; storing AX in memory addressed by [DI]
0012–0015	Register recovery
0020–0054	*WTSEND
0020–0027	Register evacuation
0028–0032	Storing first argument (source address) in Si
0033–003C	Storing second argument (destination address) in DI
003D–0041	Setting ES (destination segment) at C000H
0042–0046	Setting DS (source segment) at 5000H
0047–004B	256 word transfer
004C–0053	Register recovery
0060–0094	*TWSEND
0060–0067	Register evacuation
0068–0072	Storing first argument (source address)
0073–007C	Storing second argument (destination address)
007D–0081	setting ES at 5000H
0082–0086	setting DS at C000H
0087–008B	256 word transfer
008C–0093	register recovery

```
0001                          ****************************************************
0002                          ****************************************************
0003                          * Label definition
0004                          ****************************************************
0005   0027           TABLE  EQU   39        39 page --> 780H-7FFH
0006                          *
0007   007B                  DORG  >7B
0008   007B           RUNADR BSS   1         Execution address register
0009   007C           SDMADR BSS   1         Source Data Memory Address Register
0010   007D           DDMADR BSS   1         Destination Data Memory Address Register
0011   007E           RPTCRG BSS   1         Repeat Counter Register
0012   007F           PCTOTM BSS   1         Command Register
0013                          ****************************************************
0014                          * Interrupt Vector
0015                          ****************************************************
0016   0000                  AORG  0
0017   0000 FF80      INIT   B               Branch to Initial Set Program
       0001 0020
0018   0002 FF80      INT0   B               Branch to INT0 service routine
       0003 0040
0019   0004 FF80      INT1   B               Branch to INT1 service routine
       0005 0044
0020   0006 FF80      INT2   B               Branch to INT2 service routine
       0007 0045
0021   0018                  AORG  24
0022   0018 FF80      TINT   B               Branch to timer interrupt routine
       0019 0046
0023                          *
0024   001A FF80      RINT   B               Branch to RINT service routine
       001B 0047
0025   001C FF80      XINT   B               Branch to XINT service routine
       001D 0048
0026   001E FF80      TRAP   B               Branch to TRAP service routine
       001F 0049
0027                          ****************************************************
0028                          * Initial Set Program
0029                          ****************************************************
```

IDT   'SRVAID'

266

```
0030  0020                AORG  >20
0031  0021  CE01   INIT   DINT
0032  0021  CE04          CNFD              Configuration B0 is Data Memory
0033  0022  558C          LARP  AR4         (ARP)=AR4
0034  0023  D400          LRLK  AR4,>8000   AR4 <-- Destination Address
      0024  8000
0035  0025  CBFF          RPTK  >FF
0036  0026  FCA0          BLKP  >0,*+       (DS 8000H) <-- (PS 0000H)
      0027  0000
0037  0028  C800          LDPK  0
0038  0029  D001          LALK  >FFC1       ACCL <-- FFC1H
      002A  FFC1
0039  002B  6004          SACL  >4          IMR SET INTO ENABLE
0040                *
0041  002C  CE00   WAIT   EINT              Enable Interrupt
0042  002D  CE1F          IDLE              Wait for Interrupt
0043  002E  FF80          B     WAIT
      002F  002C
0044              *************************************************************
0045              * Interrupt Service Routine
0046              *************************************************************
0047  0040                AORG  >40
0048                *
0049  0040  C827   INT0   LDPK  TABLE
0050  0041  207F          LAC   PCTOTM      ACCL <-- (PCTOTM)
0051  0042  CE00          EINT              Enable Interrupt
0052  0043  CE25          BACC              Jump (ACCL)
0053                *
0054  0044  CE26   INT1   RET
0055  0045  CE26   INT2   RET
0056  0046  CE26   TINT   RET
0057  0047  CE26   RINT   RET
0058  0048  CE26   XINT   RET
0059  0049  CE26   TRAP   RET
```

Figure C.3 Program SRVAID.MPO

267

```
0060     ****************************************************************
0061     * Main Program
0062     ****************************************************************
0063  0050                 AORG    >50
0064     *
0065  0050 C827   DSTODS   LDPK    TABLE        DP    <-- TABLE
0066  0051 5588            LARP    AR0          ARP   <-- AR0
0067  0052 D001            LALK    SDMAPS+1     ACC   <-- SDMAPS+1
      0053 0058
0068  0054 597C            TBLW    SDMADR       (SDMAPS+1) <-- (SDMADR)
0069  0055 307D            LAR     AR0,DDMADR   AR0   <-- (DDMADR)
0070  0056 4B7E            RPT     RPTCRG       RPTC  <-- (RPTCRG)
0071  0057 FDA0   SDMAPS   BLKD    >1000,*+
      0058 1000
0072  0059 CE26            RET
0073     *
0074  005A C827   PSTODS   LDPK    TABLE        DP    <-- TABLE
0075  005B 5588            LARP    AR0          ARP   <-- AR0
0076  005C 207C            LAC     SDMADR       ACC   <-- (SDMADR)
0077  005D 307D            LAR     AR0,DDMADR   AR0   <-- (DDMADR)
0078  005E 4B7E            RPT     RPTCRG       RPTC  <-- (RPTCRG)
0079  005F 58A0            TBLR    *+
0080  0060 CE26            RET
0081     *
0082  0061 C827   DSTOPS   LDPK    TABLE        DP    <-- TABLE
0083  0062 5588            LARP    AR0          ARP   <-- AR0
0084  0063 207D            LAC     DDMADR       ACC   <-- (DDMADR)
0085  0064 307C            LAR     AR0,SDMADR   AR0   <-- (SDMADR)
0086  0065 4B7E            RPT     RPTCRG       RPTC  <-- (RPTCRG)
0087  0066 59A0            TBLW    *+
0088  0067 CE26            RET
0089     *
0090  0068 C827   RUNPRG   LDPK    TABLE        DP    <-- TABLE
0091  0069 207B            LAC     RUNADR       ACCL  <-- (RUNADR)
0092  006A CE00            EINT                 Enable Interrupt
0093  006B CE25            BACC                 Jump (ACCL)
0094     *
0095                       END
```

```
35
36                            * TWSEND
   0060    50          PUSH    AX
   0061    53          PUSH    BX
   0062    51          PUSH    CX
   0063    52          PUSH    DX
   0064    1E          PUSH    DS
   0065    06          PUSH    ES
   0066    56          PUSH    SI
   0067    57          PUSH    DI
37 0068    8B4F06      MOV     CX,06(BX)    CX <-- (BX+06)
   006B    8EC1        MOV     ES,CX        ES <-- CX
   006D    8B7704      MOV     SI,04(BX)    SI <-- (BX+04)
   0070    26          ES:
38 0071    8B34        MOV     SI,(SI)      SI <-- ES:(SI)
   0073    8B4F02      MOV     CX,02(BX)    CX <-- (BX+02)
   0076    8EC1        MOV     ES,CX        ES <-- CX
   0078    8B3F        MOV     DI,(BX)      DI <-- (BX)
   007A    26          ES:
39 007B    8B3D        MOV     DI,(DI)      DI <-- ES:(DI)
   007D    B90050      MOV     CX,5000      CX <-- 5000H
   0080    8EC1        MOV     ES,CX        ES <-- 5000H
   0082    B900C0      MOV     CX,C000      CX <-- C000H
   0085    8ED9        MOV     DS,CX        DS <-- CX
   0087    B90001      MOV     CX,0100      CX <-- 0100H
   008A    F3          REP
   008B    A5          MOVSW                ES:(DI) <-- DS:(SI)
40 008C    5F          POP     DI
   008D    5E          POP     SI
   008E    07          POP     ES
   008F    1F          POP     DS
   0090    5A          POP     DX
   0091    59          POP     CX
   0092    5B          POP     BX
   0093    58          POP     AX
   0094    CF          IRET
```

Figure C.3 (continued)

```
                        * MEMCLR
0000    50          PUSH    AX
0001    51          PUSH    CX
0002    06          PUSH    ES
0003    57          PUSH    DI
0004    B80050      MOV     AX,5000        AX <-- 5000H
0007    8E0C        MOV     ES,AX          ES <-- AX
0009    31C0        XOR     AX,AX          AX <-- 0000H
000B    89C7        MOV     DI,AX          DI <-- AX
000D    B90080      MOV     CX,8000        CX <-- 8000H
0010    F3          REP
0011    AB          STOSW                  Store AX to ES:(DI)
0012    5F          POP     DI
0013    07          POP     ES
0014    59          POP     CX
0015    58          POP     AX
0016    CF          IRET


                        * WTSEND
0020    50          PUSH    AX
0021    53          PUSH    BX
0022    51          PUSH    CX
0023    52          PUSH    DX
0024    1E          PUSH    DS
0025    06          PUSH    ES
0026    56          PUSH    SI
0027    57          PUSH    DI
0028    8B4F06      MOV     CX,06(BX)      CX <-- (BX+06)
002B    8EC1        MOV     ES,CX          ES <-- CX
002D    8B7704      MOV     SI,04(BX)      SI <-- (BX+04)
0030    26          ES:
0031    8B34        MOV     SI,(SI)        SI <-- ES:(SI)
0033    8B4F02      MOV     CX,02(BX)      CX <-- (BX+02)
0036    8EC1        MOV     ES,CX          ES <-- CX
0038    8B3F        MOV     DI,(BX)        DI <-- (BX)
003A    26          ES:
003B    8B3D        MOV     DI,(DI)        DI <-- ES:(DI)
003D    B900C0      MOV     CX,C000        CX <-- C000H
0040    8EC1        MOV     ES,CX          ES <-- CX
0042    B90050      MOV     CX,5000        CX <-- 5000H
0045    8ED9        MOV     DS,CX          DS <-- CX
0047    B90001      MOV     CX,0100        CX <-- 0100H
004A    F3          REP
004B    A5          MOVSW                  ES:(DI) <-- DS:(SI)
004C    5F          POP     DI
004D    5E          POP     SI
```

Figure C.4   Program SRVAID.MAC

```
004E      07        POP       ES
004F      1F        POP       DS
0050      5A        POP       DX
0051      59        POP       CX
0052      5B        POP       BX
0053      58        POP       AX
0054      CF        IRET
```

Figure C.4 *(continued)*

Loading a TMS 320C25 Assembler Object File

The TMS 320C25 assembler object file is a sequential file which consists of records. Each file is composed of a TAG expressed by one alpha character followed by the FIELD. The main TAGs and FIELDs are given in Table C.1.
An example of an object file is given in Table C.2.

When you load the object file, first you open the file, and obtain the record, then, partition it into the TAG and the field and store them in memories.

The flowchart for loading the object file is explained in Figure C.5.

TABLE C.1   FILE STRUCTURE

Tag	Field
K	# of program relocatable codes (4 characters) and program identifier (8 characters)
9	Absolute address (4 characters)
B	# of 16 bits (4 characters)
7	Check sum (5 characters)
F	Space, then sequential number (end of records)
:	Module name and assembling data (last record)

TABLE C.2 EXAMPLE OF OBJECT FILE

```
K0000SRVAID  90000BFF80B0020BFF80B0040BFF80B0044BFF80B0045900187F1C3F  SRVAID 1
BFF80B0046BFF80B0047BFF80B0048BFF80B00499002C0BCE01BCE04B558CBD4007F126F  SRVAID 2
B8000BCBFFBFCA0B0000BC800BD001BFFC1B6004BCE00BCE1FBFF80B002C900407F0F7F  SRVAID 3
BC827B207FBCE00BCE25BCE26BCE26BCE26BCE2690050BC827B55887F0BBF  SRVAID 4
BD001B0058B597CB307DB4B7EBFDA0B1000BCE26BCE827B5588B207CB307DB4B7E7F0FFF  SRVAID 5
B58A0BCE26BC827B5588B207DB307CB4B7EB59A0BCE26BCE827BBCE00BCE257F0D2F  SRVAID 6
       SRVAID    06-24-87  16:05:38   ASM32020 PC 1.0  85.157          SRVAID 7
```